Assessing Revolutionary and Insurgent Strategies

UNDERGROUNDS IN INSURGENT, REVOLUTIONARY, AND RESISTANCE WARFARE

SECOND EDITION

Paul J. Tompkins Jr., USASOC Project Lead

Robert Leonhard, Editor

United States Army Special Operations Command

and

The Johns Hopkins University Applied Physics Laboratory

National Security Analysis Department

This publication is a work of the United States Government in accordance with Title 17, United States Code, sections 101 and 105.

Published by:

The United States Army Special Operations Command

Fort Bragg, North Carolina

25 January 2013

Second Edition

First Edition published by Special Operations Research Office, American University, November 1963

Reproduction in whole or in part is permitted for any purpose of the United States government. Nonmateriel research on special warfare is performed in support of the requirements stated by the United States Army Special Operations Command, Department of the Army. This research is accomplished at The Johns Hopkins University Applied Physics Laboratory by the National Security Analysis Department, a nongovernmental agency operating under the supervision of the USASOC Special Programs Division, Department of the Army.

The analysis and the opinions expressed within this document are solely those of the authors and do not necessarily reflect the positions of the U.S. Army or The Johns Hopkins University Applied Physics Laboratory.

Comments correcting errors of fact and opinion, filling or indicating gaps of information, and suggesting other changes that may be appropriate should be addressed to:

United States Army Special Operations Command

G-3X, Special Programs Division

2929 Desert Storm Drive

Fort Bragg, NC 28310

All ARIS products are available from USASOC at www.soc.mil under the ARIS link.

ASSESSING REVOLUTIONARY AND INSURGENT STRATEGIES

The Assessing Revolutionary and Insurgent Strategies (ARIS) series consists of a set of case studies and research conducted for the US Army Special Operations Command by the National Security Analysis Department of The Johns Hopkins University Applied Physics Laboratory.

The purpose of the ARIS series is to produce a collection of academically rigorous yet operationally relevant research materials to develop and illustrate a common understanding of insurgency and revolution. This research, intended to form a bedrock body of knowledge for members of the Special Forces, will allow users to distill vast amounts of material from a wide array of campaigns and extract relevant lessons, thereby enabling the development of future doctrine, professional education, and training.

From its inception, ARIS has been focused on exploring historical and current revolutions and insurgencies for the purpose of identifying emerging trends in operational designs and patterns. ARIS encompasses research and studies on the general characteristics of revolutionary movements and insurgencies and examines unique adaptations by specific organizations or groups to overcome various environmental and contextual challenges.

The ARIS series follows in the tradition of research conducted by the Special Operations Research Office (SORO) of American University in the 1950s and 1960s, by adding new research to that body of work and in several instances releasing updated editions of original SORO studies.

VOLUMES IN THE ARIS SERIES

Casebook on Insurgency and Revolutionary Warfare, Volume I: 1933–1962 (Rev. Ed.)
Casebook on Insurgency and Revolutionary Warfare, Volume II: 1962–2009
Undergrounds in Insurgent, Revolutionary, and Resistance Warfare (2nd Ed.)
Human Factors Considerations of Undergrounds in Insurgencies (2nd Ed.)
Irregular Warfare Annotated Bibliography

FUTURE STUDIES

The Legal Status of Participants in Irregular Warfare
Case Studies in Insurgency and Revolutionary Warfare—Colombia (1964–2009)
Case Studies in Insurgency and Revolutionary Warfare—Sri Lanka (1976–2009)

SORO STUDIES

Case Study in Guerrilla War: Greece During World War II (pub.1961)
Case Studies in Insurgency and Revolutionary Warfare: Algeria 1954–1962 (pub. 1963)
Case Studies in Insurgency and Revolutionary Warfare: Cuba 1953–1959 (pub. 1963)
Case Study in Insurgency and Revolutionary Warfare: Guatemala 1944–1954 (pub. 1964)
Case Studies in Insurgency and Revolutionary Warfare: Vietnam 1941–1954 (pub. 1964)

About Conflict Research Group

Conflict Research Group (CRG) is a non-profit think tank based in the United Kingdom, dedicated to advancing understanding of the art and science of Unconventional Warfare. With a focus on the academic study of guerrilla warfare, revolutionary warfare, asymmetric warfare, Fourth Generation Warfare, Fifth Generation Warfare, and political unrest, CRG's work sheds light on the complexities and nuances of modern conflicts. By bringing critical and key works into print, the organization serves as a vital resource for academics, policymakers, and military professionals seeking in-depth knowledge in these specialized fields.

At the heart of CRG's mission is the belief that a comprehensive understanding of Unconventional Warfare is essential for addressing contemporary security challenges. The group's research and publications delve into historical and contemporary case studies, exploring the strategies, tactics, and implications of irregular warfare. Through this rigorous analysis, CRG contributes to the development of more effective and adaptable strategies for dealing with non-traditional threats.

One of the key aspects of CRG's work is its publishing arm, which is dedicated to bringing into print seminal works on Unconventional Warfare. The group's publications cover a wide range of topics, from historical accounts of guerrilla movements to theoretical analyses of contemporary conflict dynamics and reprints of government publications. By making these works accessible to a broader audience, CRG aims to enrich the discourse on Unconventional Warfare and contribute to the development of more nuanced and effective approaches to resolving conflicts and disrupting, degrading and defeating unconventional threats.

CRG's research is characterized by its interdisciplinary approach, drawing on insights from military history, political science, sociology, and international relations. This holistic perspective allows the organization to address the multifaceted nature of unconventional warfare, considering not only military tactics but also the granularity of the political, social, and economic dimensions of conflicts. Through this comprehensive approach, CRG provides a deeper understanding of the root causes and long-term implications of irregular warfare.

Published by Conflict Research Group.

First published by USASOC in 2013

ISBN: 978-1-925907-15-5

LETTER OF INTRODUCTION

The first time I saw the original version of the *Human Factors Considerations of Undergrounds in Insurgencies* was when it was presented to me in 1979 at the Special Forces Qualification Course, and I have reread it multiple times in the years since. A couple of years after I first read *Human Factors*, I discovered in the preface a mention of a predecessor work titled *Undergrounds in Insurgent, Revolutionary, and Resistance Warfare.* I sought out the book and it along with *Human Factors* became core references for my career in Special Forces.

The original edition of this book, and the remainder of the series produced by Special Operations Research Office (SORO),[a] constitute some of the foundational references necessary to a full understanding of Special Forces operations and doctrine. Thirty-three years later, I still use it routinely to teach and mentor new and experienced Special Forces personnel.

This book is intended as an update to the original text. Basic human nature does not change, and much has remained the same throughout the history of human political conflict. The new edition therefore includes the enduring principles and methods from the original. The sections of the book discussing techniques, causes, and methods, however, required updating to reflect the many changes in the world and experiences of Special Forces in the last fifty years.

While the need to resist perceived oppression has not changed, the ways in which oppressed societies express this need through culture have changed significantly. Whereas insurgencies prior to the publication of the first edition of this book in 1963 were predominantly Communist-inspired, modern rebellions have been inspired by a greater number of factors. This increase in the causes of insurgency began with the fall of the Soviet Union and accelerated with the attacks of September 11, 2001, on the United States.

In addition, technology has changed the world significantly over the last fifty years. Modern communications, the Internet, global positioning and navigation systems, and transportation have all introduced different dynamic pressures on how insurgencies develop and operate. As a result of the cultural and technological shifts of the past fifty years, we decided that the original series by SORO needed revision and updating. Like the original series, this update is written by sociologists, this time from The Johns Hopkins University Applied Physics Laboratory.

---

[a] They include: *Human Factors Considerations of Undergrounds in Insurgencies* and the *Casebook on Insurgency and Revolutionary Warfare*, which includes case studies on Cuba, Algeria, Guatemala, and Vietnam, and a study of guerrilla warfare in Greece during World War II.

Plenty of histories and military analyses have been written on the different revolutions that have taken place over the past fifty years, but this series—including the new, second volume of the *Casebook*—hopefully provides a useful perspective. Since rebellion is a sociopolitical issue that takes place in the human domain, a view through the nonmilitary lens broadens the aperture of learning for Special Forces soldiers.

This project would not have been possible without the support of COL Dave Maxwell, who was the USASOC G3 (my boss) and fought to get the project approved. Also I must thank Michael McCran, a long-time friend who encouraged me to continue to seek the approval of this project.

Paul J. Tompkins Jr.
USASOC Project Lead

PREFACE

Since the original publication of *Undergrounds in Insurgent, Revolutionary, and Resistance Warfare* in 1963, much has changed, and much remains relevant. The Internet, the globalization of media, the demise of Soviet Communism and the Cold War, and the rise of Islamic fundamentalism have all impacted the nature and functionality of undergrounds. The original study's observation, however, that for every guerilla fighter, there are from two to twenty-seven underground members is still true. Likewise, the report's main thesis—that the underground part of an insurgency is the *sine qua non* of all such movements—is demonstrably accurate today.

As defined in the 1963 work, undergrounds are "the clandestine elements of indigenous politico-military revolutionary organizations which attempt to illegally weaken, modify, or replace the government authority, typically through the use or threat of force." Because the art and practice of insurgency has evolved since then, almost every part of this definition deserves examination.

During the Cold War, bipolarity characterized both the world and the nature of insurgencies. Undergrounds functioned in the "illegal" realm, while the government's actions were "legal." Today, the lines of demarcation are blurred. Many insurgencies operate simultaneously in the legal, illegal, and quasi-legal domains, and governments sometimes engage in illegal activity to oppose them.

Likewise, it is hard to find the boundary between clandestine and overt operations because modern insurgencies simultaneously conduct both. Part of the reason for this blurring is a change in the paradigm of how insurgencies succeed or fail. The classic and simplistic way of thinking about it is to imagine an insurgency gaining strength and momentum over a long campaign and finally seizing control of the government in the mold of the Maoist takeover of China. Today it is more likely that the successful insurgency will gain some level of legitimate, open political acknowledgment while simultaneously continuing in quasi-legal and illegal activity. Power-sharing arrangements are more common today than revolutionaries overthrowing the government.

A modern insurgency thus can be thought of as including four components, instead of just the three described in the original work: the underground, the guerilla component, the auxiliary, and the *public component*. At the start of an insurgency, the underground might be the only active sphere. As time goes on, auxiliary and guerilla contingents begin to grow and operate. Eventually, pursuant to a political agreement, the insurgency can begin to operate in the public political process. If successful there, the entire movement might at some point become public—e.g., when they become the sole legitimate government or are otherwise fully integrated into the political process. Alternately, failure to achieve their goals in the political process might cause the movement to revert to clandestine, illegal, and violent operations. Thus, the four components—the underground, the guerilla component, the auxiliary, and the public component—exist in a dynamic and constantly changing relationship with each other.

Hence, this study will look at the public component as a requisite part of modern insurgencies, separate from but closely integrated with the underground. The public component is different from the underground in that it is *not* clandestine and most often a legal entity. At the same time, it overlaps with the underground in that the latter's functionality includes the management of propaganda and communications in general.

Likewise, our understanding of modern insurgencies requires further elucidation concerning the nature of the guerilla component. The dictionary definition of the word links it to irregular warfare, and for most of modern history since the nineteenth century, this linkage is accurate. Because the Maoist model of insurgency features an extended phase of clandestine operations aimed at garnering strength, guerilla activities conformed to the irregular label easily: i.e., insurgent soldiers employed unconventional tactics and organization against government forces. Since the end of World War II, however, the definition of the guerilla component has migrated. Modern insurgencies continue to prosecute irregular warfare, but some also seek to build conventional capability. This phenomenon was apparent in the struggles of the Viet Cong during the French and American occupations of Vietnam and is likewise a part of Hizbollah's strategy in Lebanon today. Hence, this study will use both "armed component" and "guerilla component," depending on the context, to indicate that modern insurgencies often use both regular and irregular organization and tactics.

The original study offered a germane and well-considered discussion of the definition of undergrounds. It concluded that criminal organizations, although by definition clandestine and illegal, fell nevertheless outside the definition of undergrounds for their lack of a political-military organization and an objective of opposing the government. As with other aspects of the definition, this conclusion has lost strength. Indeed, modern insurgencies now almost universally include both criminal activity and some form of alliance or cooperation with criminal networks. The line between the strictly ideological insurgent and the strictly criminal drug lord has blurred. Pablo Escobar—the notorious Colombian drug lord killed in 1993—evolved from a criminal gang leader into a quasi-political figure who dabbled in social activism, for example, and the FARC in Colombia found itself inextricably associated with the criminal drug trade.

Finally, the original study of undergrounds discussed the most relevant functions, but the present study includes a look at the function of leadership specifically. The leaders of insurgent, resistance, and revolutionary movements often create or emerge from the underground. Underground leaders provide strategic and tactical direction, organization, and the ideology of the movement. They perform these functions within the unique and compelling context of their country, culture, and political economy. How they manage the often conflicting trends that define the framework of their insurgent movements in large measure determines ultimate success and failure.

The purpose of this book is to educate the student and practitioner of insurgency and counterinsurgency. Just as a doctor must study the anatomy of the human body in order to advance in the profession of medicine, so the insurgent or counterinsurgent must understand the anatomy of the insurgent movement. The doctor does not confine himself to treating visible symptoms, but instead looks for causality within the body. In a similar manner, counterinsurgency does not aim solely at the guerilla component of the enemy but must also operate against the underground that sustains the more visible aspects of the insurgency. This book examines the anatomy of undergrounds in various insurgencies of recent history. Our goal is to continue the groundbreaking work performed in the original study and update it with insights from the post-Cold War world.

Primary source material for this book comes from the Tier I and Tier II Case Studies written as part of the Assessing Revolutionary and Insurgent Strategies project. Hence, these case studies should be used as companion documents for this study.

Robert Leonhard

TABLE OF CONTENTS

LIST OF ILLUSTRATIONS

CHAPTER 1.

LEADERSHIP AND ORGANIZATION

CHAPTER CONTENTS

Robert Leonhard and Jerome M. Conley

INTRODUCTION

It is at once obvious that the most fundamental element of an insurgent underground is its organization and the leadership that guides it. TC 18-01 notes that insurgencies require "leadership to provide vision, direction, guidance, coordination, and organizational coherence."[1] Stated differently, leaders form movements, provide them with an ideological direction, recruit members, and organize these members to achieve the movement's goals. The movement's goals—and the overall operating environment—determine the organization of the movement. Hence, in order to understand an insurgent underground, a necessary first step is to comprehend the nature of its organization and its leadership. The companion to this work, the second edition of *Human Factors Considerations of Undergrounds in Insurgencies,* deals with how insurgent leaders emerge and focuses on their psychology and sociological development. This work, conversely, looks at the structure and functions of organizations and the functionality of leadership.

The primary factor that determines the organization of an insurgency and its underground is the goal of the movement. In large part, this goal is defined by the ideology of the movement, so this chapter will begin by discussing the role of ideology. It should be noted, however, that many insurgencies (with the possible exception of Communist movements) are started by individuals who lack prior experience with designing and leading a movement, resulting in an initial organizational structure that will likely evolve as these leaders gain experience and as the resource and security aspects of the movement change.[2]

IDEOLOGY

The crucial role of ideology in an insurgent or resistance movement cannot be overstated. An ideology grows out of discontent with the status quo; it is the intangible idea that gives rise to acts of defiance and rebellion. Ideology also plays a dual role in an insurgency in that it serves as the basis for recruitment and it illuminates strategic direction.

These observations seem obvious. What is not so obvious is the decisive role leaders assume as they cultivate, develop, and evolve ideology during the course of an insurgency. Their decisions and motivations regarding the specific ideological principles of the movement—political, economic, cultural, or religious—are anything but academic. Rather, they serve to characterize the movement's ability to appeal to the masses, and they both energize and—in many cases—constrain the progress of the movement's strategy.

A key principle for understanding modern insurgent, resistance, and revolutionary movements is that *ideology evolves*. It is rare for a movement's ideology to remain unchanged throughout the course of the struggle. An insurgency's ideology moves along a spectrum of exclusivity and inclusivity as leaders stake out the movement's position on politics, religion, social justice, etc. *Exclusive* ideologies aim at energizing a targeted sector of the population, helping the members of that population to define themselves in relation to the foes they oppose. *Inclusive* ideologies, conversely, seek to unify various groups and encourage them to coalesce around the insurgency's main goal. Exclusive ideologies have the advantage of facilitating strategic focus because they tend to embrace very specific and dramatic goals. The disadvantage of exclusivity, however, is that the ideology does not appeal to a broad sector of the population. Inclusive ideologies tend to embrace large portions of the population, but they suffer from multiple, vague, and often conflicting goals that make strategic focus problematic. See Figure 1-1.

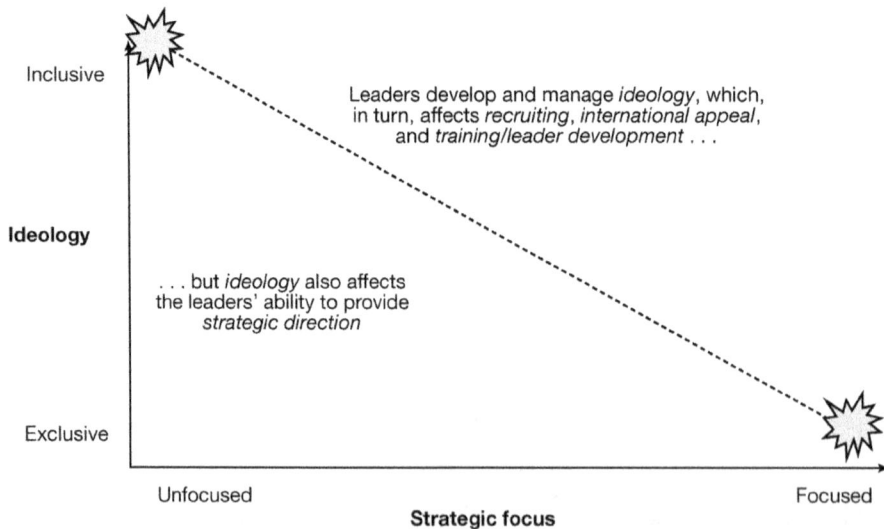

Figure 1-1. Ideology and strategic focus.

Yasser Arafat's leadership of Fatah and, later, the Palestine Liberation Organization (PLO) illustrates both the strengths and weaknesses of an inclusive approach to ideology. The most fundamental goal of the PLO was the destruction of Israel so that displaced Palestinians could return to their homeland. Beyond that single point of congruity, however, there were many competing ideas within the movement. Pan-Arabists viewed the unification of Arabs as the key to achieving the overall goal (the destruction of Israel) and agitated for the PLO to subordinate itself to the Arab leaders of surrounding states—primarily Egypt's Gamal Abdel Nasser. Communists insisted that the entire Palestinian conflict was a manifestation of the universal struggle against capitalism and imperialism. Islamists interpreted the conflict in theological and eschatological terms and pushed for religious revival as the means to victory. Confronted with these and other disparate ideas, Arafat welcomed them all, at least initially. He deftly courted the various subgroups and factions that would ultimately compose the PLO but nurtured a singular focus on destroying the hated Israeli foe. Arafat's inclusive ideology had utility because it allowed the PLO to adapt to these new, vigorous ideas that were emerging in the Arab world. The ideology facilitated the continued relevance of the PLO, whereas a stricter approach would likely have alienated potential partners and splintered the organization into nonexistence. In short, an inclusive attitude led to the growth and sustainment of the movement.

The drawbacks, however, were equally obvious. Arafat was under constant attack from leaders within his own organization. To them, he was not Arab enough, socialist enough, or Islamic enough. Not only did this inherent disunity foil his attempts to cement his control of the organization, but it also led to various subgroups "hijacking" PLO strategy. As factions sought to dominate both the headlines and the parent organization, leaders would sometimes engage in spectacular acts of terror or other violence. Arafat often found himself racing to keep up with events and trying to rein in recalcitrant colleagues. In this he was never fully successful.

An example of the opposite, exclusive approach is found in the case of the Lebanese Hizbollah insurgency. Having splintered from the Amal Movement in the mid-1980s, Hizbollah leaders issued a manifesto, referred to as the "Open Letter," in 1985. In the manifesto they described their ideology in crystal clear terms: they were Islamic, strongly allied with Iran, and utterly determined to reject Lebanese politics. This type of strong rhetoric appealed to their chief constituents—the impoverished, war-weary, and practically disenfranchised Shiites in the south. Bolstered by the theological convictions and edicts of a respected cleric, Sayyid Muhammad Fadlallah, the energetic young insurgency attracted many recruits.

But in the early 1990s, a key crossroad appeared when Hizbollah leaders had to wrestle with the question of entering the very political arena that they had previously spurned. Hassan Nasrallah, the leader of Hizbollah, made the fateful decision to change the group's ideology in this regard. Hizbollah thereafter ran for seats in the Lebanese parliament and captured an influential sector of the government. Nasrallah reckoned that the time had come in which the insurgency, if it was to continue helping its constituents, would have to develop a legal, political presence in Beirut. His decision led to the inevitable and predictable loss of some supporters whose ideological rigidity did not facilitate such maneuvers, but in the end, Hizbollah benefited from the change. Hence, in contrast to the PLO, Hizbollah began with a very exclusive ideology but gradually expanded its appeal and influence by changing to have a more inclusive stance.

Sendero Luminoso (Shining Path), the Maoist insurgency in Peru, developed an exclusive ideology rooted in the ethnic and class conflict within the country. The movement allied itself with the largely disenfranchised and impoverished indigenous population, and that ideology garnered immediate and widespread support within the local communities of the Peruvian highlands. However, this ideology did not resonate as strongly with other communities in the coastal plain or in the urban communities of Lima and other Peruvian cities. Oddly, Abimael Guzman, the leader of Sendero, chose to not reach out to other constituencies or to potential external sources of support. Instead, his ideology, and that of his followers, remained focused on the centrality of Guzman and the Maoist model he endorsed within Peru for addressing ethnic and class conflict. Consequently, the exclusivity of the Shining Path ideology served as a brake on the movement's progress, just as it initially helped to accelerate the movement's development in the early stages.[3]

Finally, and not surprisingly, the complete absence of a strong, unifying ideology (whether inclusive or exclusive) can severely limit the development, growth, and sustainment of an insurgency, especially when this ideological void is filled by personal motivation and ambitions. Insurgencies such as the Fuerzas Armadas Revolucionarias de Colombia (Revolutionary Armed Forces of Colombia, or FARC) in Colombia, the Revolutionary United Front (RUF) in Sierra Leone, and the Movement for the Emancipation of the Niger Delta (MEND) in Nigeria all started with pseudo-ideological foundations related to government repression and government control of resources but soon deteriorated into struggles between local insurgent leaders and their control over drug crops, diamond fields, and oil bunkering operations (respectively). In this environment, where individual aspirations usurp any strategic ideology, organizational cohesion quickly deteriorates.

These examples are repeated in many modern insurgencies today because ideology remains closely wedded to a movement's ability to recruit, build partnerships, and act decisively. Ideology can be a strategic tool for flexible, adaptive leaders, or it can be a straitjacket that inhibits team building and leads to ineffective strategy.

ORGANIZATION

An underground must develop an effective organization in order to carry out its operational missions. An effective organization in turn requires the underground to perform certain essential "housekeeping" administrative functions as well as operational functions. In discussing these functions in this chapter, we will enumerate the major problems likely to be faced by most undergrounds and describe alternate ways in which the problems have been handled through organization.

There are no set or standardized ways to accomplish a given function. The choice and effectiveness of a given technique depends largely upon the resources available to the movement and the countermeasures used by the security force. As undergrounds adopt certain practices, the security forces invariably develop countermeasures that undermine the effectiveness of these practices. Consequently, both governments and undergrounds are constantly changing techniques and developing new ones. Moreover, the availability of resources governs the underground's choice of techniques and determines how flexible it can be in responding to government actions.

Insurgencies in the modern world populate a spectrum of sophistication from jungle tribes employing drums for communication to Internet-savvy urban operatives conversant with the latest trends in science and technology. In less-developed regions, the absence of legacy communication systems and the availability of significant funding from trafficking in gemstones, weapons, oil, humans, narcotics, etc., can result in insurgent groups having the ability to leap-frog over the technical capabilities of the security forces through the purchase of sophisticated equipment through either legal or black markets. Thus, the available means, participants' education level, and the nature of the government security apparatus combine to frame the underground's organizational and communications strategy.

Finally, there is no single organizational structure that is best suited for an underground under all conditions. A number of major factors must be considered by those responsible for organizing the underground, and on the basis of these considerations, an organizational pattern is used that will best provide for successful accomplishment of underground functions. As stated above, however, competing internal interests as well as lack of experience can often result in an initial organizational process that is ad hoc or ill-conceived. A discussion of the major factors that must be considered during organizational planning is presented below and is followed by a description of the patterns of organization that reflect critical insurgency principles.

Factors Determining Organizational Structure

Predominant Strategy

Resistance and revolutionary movements operate on both political and military fronts. The choice of which front an underground will emphasize depends on the likelihood of success in one area rather than another and on unexpected events that compel the underground to revise its predominant strategy. It may be that underground infiltration of key positions in both the government and the military is blocked, as is usually the case in resistance situations and often the case in revolutionary ones. When it is impossible to seize the instruments of control, the underground chooses the predominantly political strategy of trying to weaken the government's effectiveness by sabotage and by subverting the people's support of the government.

An underground may shift from a predominantly political strategy to one that is predominantly military because it is losing popular political support. The leaders may feel that unless they act immediately and aggressively, they may lose their last opportunity to seize power. In post-World War II Philippines, the Communists were restrained from continuing their activities on the political front when they were denied access to seats in the legislature. To offset the drop in popular support created by this curtailment of their activities, they consequently turned to military action. A rapid shift to a predominantly military strategy can also occur when popular sentiment against the existing government becomes so strong that the insurgency's leadership feels that they can use this popular discontent to quickly seize power, as the Communists in Greece believed after World War II.

Having decided to shift primary emphasis from either the political or military front to the other, the entire underground may require major reorganization. General Vo Nguyen Giap reported that in Indochina, the transition from political to armed struggle caused a great change in the principles of organization and work. During the Malayan Emergency in the late 1940s, Communist leaders shifted from political efforts, through the organization of united fronts and infiltration of unions, to the militarization phase of terrorism and guerrilla warfare. After this failed, they turned back to political activity. With each shift in strategy, various organizational units had to be organized, disbanded, or reorganized, all at no small cost to the movement. These units included escape and evasion nets for both guerrillas and underground; intelligence nets in urban areas, as well as throughout the countryside; secret supply depots and supply routes; and recruiting teams for both underground and military units. In addition, attempts to transition from military to political strategies can often fail because the military success of insurgent organizations often relies on decentralized control and the independence of local field commanders, but this same independence can make it impossible to create a unified, centralized organization that is critical for the transition to a political strategy. Recent examples include the campaigns by the RUF in Sierra Leone and the MEND in Nigeria.

The expected duration of the underground also has a very important effect upon its organization and its activities. Movements that are organized for a very short period of time do not usually have a complex organization, elaborate security procedures, or selective membership. Underground movements that expect to exist for several years often develop more complicated organizations and undertake a wider range of activities.

The important point is that different predominant strategies will give different emphases to underground functions, and performance of these various functions, in turn, requires different organizational structures.

Origins and Leadership

Undergrounds usually develop within existing social, political, or military organizations. The underground organizer must rely on past relationships so he will not be betrayed and on the informal communication channels of organizations with which he is familiar. As a result, many undergrounds have emerged from outlawed political parties, labor organizations, civic clubs, or disbanded military organizations. This was a critical element of the Euskadi Ta Askatasuna (Basque Homeland and Freedom, or ETA) organizational structure in Spain as it was built around the Basque institution of the "cuadrilla," a group of friends of roughly the same age who spent most of their time together drinking, sharing meals, and mountain climbing, with group ties often stronger than family ties.[4] Existing organizations can also provide the necessary structure and relationships for recruiting members for an insurgency, which was the deliberate strategy of the Hutu leadership in Rwanda when it organized tens of thousands of recently unemployed young males into soccer clubs that would eventually become the operational elements of their genocide against the Tutsis in 1994.[5]

Historically, the organizational character of the underground movement has differed according to whether the movement has its beginning in political or military organizations and whether it was led by a political or military leader. Organizations controlled by politicians are frequently found in revolutionary warfare. Also, many of the European underground organizations of World War II developed from prewar political parties and were directed by political leaders. Consequently, the organization reflected the influence of the political leader.

On the other hand, although many resistance groups may organize spontaneously, national groups are usually sponsored and supported by external conventional military forces. Because it is usually through this military assistance that the movement gains its stature, the choice of underground leaders is influenced by their knowledge of strategy, tactics, and military affairs, and their activities are oriented toward providing aid to conventional forces. In Yugoslavia, for example, the Allied forces withdrew support from Dragoljub Mihailovic and recognized and supported Marshal Josip Broz Tito during World War II because they felt he was carrying on aggressive action against the Germans and would thereby be of greater assistance to the military mission. Conversely, if the external sponsor no longer considered the insurgency to be of strategic importance to the sponsor's overall goals, it may rapidly reduce its support to the insurgency, as was the case with the Frente Farabundo Martí para la Liberación Nacional (Farabundo Marti National Liberation Front, or FMLN) in El Salvador when the critical flow of arms, money, and training for the Soviet Union was curtailed in the late 1980s, effectively dooming the insurgency.[6] Some movements have both political and military leaders who function independently of each other. In this form of organization, the underground operates politically in the populated areas, while the guerrilla forces are led by a military man in the rural areas. This occurred in the anti-Fascist resistance in Italy during World War II and in Morocco during the fight for independence (1953–1955).

Types of Organization

Organizational structure varies with the organizational theories of the resistance or the revolutionary leaders.

Mass Organization

When leaders conclude that a large number of people are necessary to overcome the power of the governing authority and its instruments of force, they may opt for mass organization. Membership is open to anyone who wishes to join, and the objective is to recruit as many people as possible. For the Liberation Tigers of Tamil Eelam (LTTE) in Sri Lanka, the attempt to mobilize the entire population in the territory it controlled led it to create in 1999 a policy for a "Universal People's Militia" that would impose military training on anyone over the age of fifteen.[7]

One disadvantage of this organizational structure is the loose security measures associated with it. The members are usually not practiced in security precautions, and the identities of underground members are easily obtained through loose talk and careless, overt actions. However, organizations of this type have managed to minimize the threat of informers primarily through the public sympathy for the movement and through the use of terrorism, as was the case of the Provisional Irish Republican Army (PIRA) and its practice of "knee-capping" informants. In the early days of the Organisation de l'Armée Secrète (Organization of the Secret Army/Secret Armed Organization, or OAS) in Algeria, General Raoul Salan moved about openly, even while the French government was seeking him and had posted a reward for him; the populace was terrorized into remaining silent. Another disadvantage of this type of organizational structure is in its command and control structure. It is difficult to obtain concerted action against the governing authority usually because of the lack of training and discipline of members.

Elite Organization

The theory here is that a small elite organization can make up in skill and discipline what it lacks in size and that at the proper moment, a small militant group can accomplish more in one blow than a large mass organization can accomplish over a prolonged period of time. The membership in a movement such as this is small, and each individual is carefully screened and tested before he is permitted to join. Once a member, he is subjected to intensive training and discipline to develop the skills necessary for clandestine work. This type of organization usually works toward a coup d'etat, or a revolution from the top. In a police state, where the mechanisms of internal security are extensive, this is the most common form of underground. An elite organization generally must have some condition of internal confusion, such as the assassination of the head of state or rivalry between several major political factions, before it can strike with any hope of winning. The disadvantage of this type of organization is that it must remain relatively inactive while waiting for the proper moment, and inactivity usually works against a movement because its members may lose their enthusiasm.

Elite-Front Organization

Communist insurgents have historically worked from this type of organizational theory. Recruitment is very selective, and the party itself does not expand rapidly. Instead, a "front" organization is created or an existing organization that claims to seek some popular objective such as liberation or independence is infiltrated. Within the front movement, military and civilian organizations are established. Because the Communists organize these groups, they are usually in leadership positions. If the movement fails, the Communist underground is not damaged either organizationally or by reputation because it is the front group and not the Communists who lose the insurgency. On the other hand, if they are successful, the Communists are in firm control of the revolutionary organization.

Conflicting Needs of Security and Expansion

An underground operates in a hostile environment in which government forces attempt to seek it out and destroy it. In order to survive, and at the same time achieve its ultimate objectives, an underground must adapt to two major factors, government countermeasures and its own successes and failures. Changes in its organization and its size depend on the effectiveness and size of the government security forces. As the government becomes more effective, the underground must emphasize security—and this usually means smaller and fewer organizational units.

An underground may, after a few spectacular successes, find itself deluged with new recruits. If it fails to expand its organization quickly by relaxing security measures, it may pass up an opportunity to seize power and attain its objectives. On the other hand, if it recruits unreliable persons, a serious security leak may enable the security forces to destroy the entire organization. Alternately, rapid expansion may provide government security forces with the opportunity to infiltrate the underground.

Thus, undergrounds face a dilemma of conflicting goals. In order to achieve their objectives, they must be expansive and aggressive; in order to survive, they must take precautions and emphasize security. Leaders must juggle organizational patterns and size so as to achieve an optimum balance between the need to expand and the need to maintain security.

Command and Control

To be effective, an underground needs concerted action. Unless it can establish a centralized command, an underground's activities occur in a haphazard manner and may lose much of their cumulative effect (e.g., units may attack the same targets and draw government attention to each other). However, for security reasons, decentralization of activities is most desirable. Therefore, in practice, leaders develop a balance between centralized and decentralized command and control.

Because a command echelon cannot direct each subordinate unit, it must rely on mission orders; that is, the central command issues orders describing the tactical objective and recommends activities that it believes can best accomplish the objective. Each of the subordinate units, which must place a premium on survival, can devise its own plan for carrying out the orders. Consequently, the subordinate units usually have the authority to make independent decisions on local issues and to operate autonomously with only general direction and guidance from the centralized command. When special assignments are given, the central command may send a special representative to the subordinate underground unit to supervise the attack directly. In this case, the special representative, being responsible to the central command, is usually placed in charge of the unit.

The central command may not know precisely how many members belong to the subordinate units of the organization. It may test the strength of the organization by calling demonstrations, strikes, or other trial actions. Such tests provide the central command with some estimate of underground strength and ability to react without jeopardizing the identity of its members. The tests can also help the organization determine the length of time necessary to mobilize its units through the clandestine communications net, the approximate number who can be reached, and the number who ultimately participate.

The underground usually specifies in great detail the transfer of authority as well as the procedure for reestablishing the chain of command in case a leader is captured. This is to ensure that the capture of an important leader will not necessarily lead to the collapse of the organization or an interruption of its activities.

Centralization of Administrative Functions

Undergrounds that function for a prolonged period of time generally centralize many activities in the central command so that subordinate units may receive services that they could not ordinarily provide for themselves. Activities such as the production of false documents, the collection of funds, the purchase of supplies, the analysis of intelligence information, and security checks on new recruits may be better performed by a central agency. These centralized activities can best be performed outside of the country, in areas in which government security measures are lax, or in a place of sanctuary.

Here members can meet openly and discuss plans and procedures without fear of being captured or of having records fall into the hands of the security forces. In Algeria, the internal command was buttressed by an external political-military command that was usually in Tunisia. In the Philippines, the Communists had not only an internal underground command, the "politburo-in" (in Manila), but also a "politburo-out," safely located in guerrilla-held territory where much of the collating of intelligence and planning took place. The Ushtia Çlirimtare e Kosovës (Kosovo Liberation Army or KLA) in Kosovo relied heavily on administrative support that it received from its diaspora in Germany and elsewhere. During World War II, much of this centralized activity was conducted for European governments by their governments-in-exile, which were located in England.

Decentralization of Units

The basic underground unit is the cell. The cell usually has from three to seven members, one of whom is appointed cell leader and is responsible for making assignments and checking to see that they are carried out. As the underground recruits more and more people, the cells are not expanded—rather, new cells are created.

The cell may be composed of persons who live in a particular vicinity or who work in the same occupation. Often, however, the individual members do not know the places of residence or the real names of their fellow members, and they meet only at prearranged times. If the cell operates as an intelligence unit, its members may never come in contact with each other. The agent usually gathers information and transmits it to the cell leader through a courier, through a mail drop, or through clandestine use of computer networks (e.g., e-mail, social media, coded websites). The cell leader may have several agents, but the agents never contact each other and only contact the cell leader through intermediaries. Lateral communications and coordination with other cells or with guerrilla forces are also carried out in this manner. In this way, if one unit is compromised, its members cannot inform on their superiors or other lateral units.

To reduce the possibility of its members being discovered, the underground disperses its cells over widely separated geographic areas and groups. This extends the government security forces so that they cannot concentrate on any single area or group. For example, the underground attempts to gain as wide a representation as possible among various ethnic and interest groups. The Malayan Communist Party was easy prey for security forces partly because it was almost entirely composed of indigenous Chinese. The OAS in Algeria was composed of Europeans who numbered only one out of every ten inhabitants. Even more crucial for the OAS, it was centered in three cities, thus allowing the French to concentrate their forces. The Frente de Liberación Nacional (National Liberation Front, or FLN), on the other hand, was made up of Muslims, who constituted 90 percent of the population and were distributed throughout the country.

An underground is usually organized into territorial units. The size of the units depends on the density of the population and the number in the underground. Each territorial unit is subdivided into districts and finally into cells. Within each of these geographical districts, and on each level of organization, different functions and groups responsible for those functions are represented. During the early 1990s, the FARC in Colombia organized into seven operational regions: Northern (Caribbean), Northwestern (bordering Panama), Middle Magdalena (along Venezuelan border), Central, Eastern, Western, and Southern; and each region had a military "Block" associated with it.[8] In addition, undergrounds have found it convenient to organize units within existing occupational activities such as railroad workers unions. Thus, the underground is organized by territorial units, such as state, county, city, and cell, as well as by professional and occupational groups, which transcend traditional boundaries.

The underground organization and many of its activities are based upon a "failsafe" principle; that is, it is organized so that if one element fails, the consequences for the total organization will be few. Almost all clandestine organizations that are susceptible to compromise by security forces have parallel organizational units and networks of units. In every case, the underground attempts to have a backup unit that can perform the same duties as the primary unit if the latter is compromised. It usually takes a long time to establish a unit or net, and the underground must plan for contingencies such as the compromise of the primary unit or increased government security measures. Thus, the organizational expansion of undergrounds is usually in a lateral direction by duplicating units and functions. The decentralization extends to all functions.

The underground usually does not jeopardize intelligence units by demanding that they perform sabotage as well, because sabotage operations may draw attention to individuals and compromise their usefulness as intelligence agents.

An underground organization that provides an excellent example of all of these principles and which served as a model for other insurgent movements is the Provisional IRA (PIRA). In response to increasingly effective countermeasures by the British in 1972 and swelling numbers of volunteers, the PIRA relied on a cellular structure for the organization to enhance both security and operational effectiveness. The basic "cell" of the organization was the Active Service Unit (ASU), which carried out the military operations of the PIRA. Usually containing four volunteers and one operations commander,[a] the ASUs could be supported by an intelligence officer or education officer if mission planning or training were required. Each ASU was responsible for the bulk of the operational expenses as well as their safe houses and transportation. The operations commander of an ASU usually knew the identity of only one higher commander in his organization, the Brigade Adjutant, and the Brigade Adjutant took his orders from the geographical or central command above him, with the PIRA having two overall commands, the Northern and Southern Commands.[b] The senior echelons of the PIRA consisted of the General Army Convention, the Army Executive, the Army Council, and the General Headquarters (GHQ). Officially located in Dublin, although many staff members were located elsewhere, the GHQ was composed of approximately fifty to sixty people who provided the centralized functions of the PIRA, to include Finance, Security, Quartermaster, Operations, Foreign Operations, Training, Engineering, Intelligence, Education, and Publicity. The Army Council was a seven-member panel that usually included the GHQ Chief of Staff, the Adjutant General, the Quartermaster, the Head of Intelligence, the Head of Publicity, and the Head of Finance. Meeting at least once a month, the Army Council approved all strategy, policy, and major actions of the PIRA. The Army Executive was a board of twelve very senior and experienced IRA veterans who met every six months to review the activities of the Army Council and who elected the seven members of the Army Council (members could not belong to both the Army Council and the Army Executive). According to the Constitution of the PIRA, the supreme authority of the movement was the General Army Convention, which was a large body of delegates that voted on the most important issues for the PIRA, such as the declarations of peace as well as the membership of the Army Executive. The Army Convention contained 100–200 active volunteers, representatives for the imprisoned, all members of the Army Council, and staff members from the local brigades, the two commands, and the GHQ. Although the Constitution allowed for the Army Convention to meet every two years, security concerns often required delays in holding such a large gathering of members.[9]

---

[a] Part-time members were often men and women who held normal jobs in the community and who would participate in PIRA operations on the weekends or after work hours. A few full-time members were paid by the PIRA with a weekly stipend and received additional support from the community in the form of donations, food, and clothing.

[b] The Northern Command was the main area of operations for the PIRA and encompassed all of Northern Ireland as well as five bordering counties. The Southern Command covered the rest of Ireland and mainly provided logistical support to the Northern Command, to include training, funding, safe houses, and the storage and movement of arms.

CONCLUSION

The leadership and organization of the underground is perhaps the most defining feature of an insurgent or resistance movement. Analysis of the underground must focus on the movement's ideology, organizational pattern, and leadership style. The most successful insurgencies demonstrate resilient organizations whose leaders possess the flexibility and vision to evolve the movement's controlling ideology, organization, and methods of command and control.

ENDNOTES

[1] TC 18-01, *Special Forces Unconventional Warfare* (Washington, DC: Headquarters, Department of the Army, 2011), 2-3, 2-7.

[2] ST 31-202, *The Underground* (Fort Bragg, NC: United States Army Institute for Military Assistance, 1978), 21. In addition to Communist movements, recent examples of the "colored revolutions" demonstrate a fairly high degree of standardization in the organization of these movements due to the common involvement of key individuals and sponsors.

[3] Ron Buikema and Matt Burger, "Sendero Luminoso," in *Casebook on Insurgency and Revolutionary Warfare, Volume II: 1962–2009*, ed. Charles Crossett (Laurel, MD: The Johns Hopkins University Applied Physics Laboratory, 2009), 58–59.

[4] Chuck Crossett and Dru Daubon, "ETA: Euskadi Ta Askatasuna" (working paper, The Johns Hopkins University Applied Physics Laboratory, Laurel, MD, 2009).

[5] Bryan Gervais, "Hutu-Tutsi Genocides," in *Casebook on Insurgency and Revolutionary Warfare, Volume II: 1962–2009*, ed. Charles Crossett (Laurel, MD: The Johns Hopkins University Applied Physics Laboratory, 2009).

[6] Ron Buikema and Matt Burger, "Farabundo Marti Frente Papa La Liberacion Nacional (FMLN)," in *Casebook on Insurgency and Revolutionary Warfare, Volume II: 1962–2009*, ed. Charles Crossett (Laurel, MD: The Johns Hopkins University Applied Physics Laboratory, 2009).

[7] Maegen Nix and Shana Marshall, "Liberation Tigers of Tamil Eelam (LTTE)," in *Casebook on Insurgency and Revolutionary Warfare, Volume II: 1962–2009*, ed. Charles Crossett (Laurel, MD: The Johns Hopkins University Applied Physics Laboratory, 2009).

[8] Ron Buikema and Matt Burger, "Colombian Insurgencies (FARC/ELN)," in *Casebook on Insurgency and Revolutionary Warfare, Volume II: 1962–2009*, ed. Charles Crossett (Laurel, MD: The Johns Hopkins University Applied Physics Laboratory, 2009).

[9] Chuck Crossett and Summer Newton, "The Provisional Irish Republican Army: 1969–2001," in *Casebook on Insurgency and Revolutionary Warfare, Volume II: 1962–2009*, ed. Charles Crossett (Laurel, MD: The Johns Hopkins University Applied Physics Laboratory, 2009).

CHAPTER 2.

RECRUITING

CHAPTER CONTENTS

Robert Leonhard and Jerome M. Conley

INTRODUCTION

Insurgencies and revolutionary movements comprise people; hence, recruiting and retention of personnel is an essential component of any sustained movement. Without a steady and reliable source of people to propel the movement onward, revolutionary leaders become ineffectual visionaries. Modern insurgencies therefore focus on finding, investigating, contacting, and assimilating people from the start of the movement all the way through to its success or demise.

Recruiting activities unfold across a set of four parameters that frame the insurgency: phasing, recruiting to task, target populations, and techniques. This chapter explains the four basic parameters of recruiting operations with the objective of educating the insurgent who is leading a resistance movement and the counterinsurgent who must seek to interdict such activities. Overall, the struggle between the insurgent and the counterinsurgent is a contest for which side can win over the population. By understanding insurgent recruiting practices, the counterinsurgent can better resist these practices, interfere with them, and/or stop them outright.

The first three parameters (phasing, recruiting to task, and target populations) establish the rationale for the techniques (the fourth parameter) that are employed in various recruiting operations. For example, an insurgency that is in its nascent stages might take an extended period of time to investigate and cultivate a relationship with a potential front leader in an area of a country targeted for expansion of the movement. Because the objective is to recruit a long-term senior leader, security is of the highest priority, and insurgent operatives would seek to develop personal ties and ideological understanding with the target. Later, the same insurgency seeking to flesh out a guerilla army might resort to mass recruiting operations built around monetary incentive with less attention paid to the security risk created by recruiting individual foot soldiers.

One general recruiting principle applies across all four parameters: recruiting operations are closely linked to insurgent ideology and objectives, and they are constrained by security risks. The strategic objectives chosen by insurgent leaders (e.g., overthrow of a government, spreading of Communism throughout a region, purging of religious opponents) derive from and help to shape the movement's ideology. That ideology, in turn, provides the fundamental basis of appeal and motivation to the targeted population. A revolutionary Islamic movement, for example, must adopt and proselytize according to a congruent religious ideology. Recruiting operations, then, will tend toward reaching a Muslim population that is open to such ideology.

A key factor in an insurgency's success, however, is strategic and ideological flexibility. As a movement progresses and expands—and especially as it begins to enjoy political success—there is often the need to evolve the ideology in such a manner as to broaden the movement's appeal and potential recruitment base. As an example, Hizbollah began as a movement with a very strict and exclusive Shiite Muslim ideology and was therefore able to reach the disaffected Shiites in southern Lebanon. As the movement grew, however, leaders saw the need and opportunity to reach out to selected groups of Maronite Christians, Druze, and Sunni Muslims as well. Consequently, the verbiage of their ideology evolved to describe a desired political economy that could accommodate religious freedom, at least to a degree.[1]

Recruiting operations must also unfold within a risky security environment. Insurgencies almost always begin as illegal movements. Exposure of key leaders or bases invites instant neutralization by government forces. Hence, especially in the early stages of an insurgency, leaders must take pains to mitigate risk through careful intelligence assessments of potential candidates prior to initiating the recruiting process, especially when seeking senior and mid-level leaders. High-visibility opposition members may serve as charismatic leaders within an underground, but because they are likely to be under surveillance as a result of their vocal opposition, they may best serve the movement as guerrilla leaders or leaders within the government in exile. Security risks attend the later stages of insurgent recruiting as well, because as operations expand, it becomes easier for counterinsurgents to infiltrate the organization, especially at the lower levels. Thus, unlike recruiting operations in a business corporation, insurgent recruiting is always constrained by security risks.

This chapter, as part of a book looking at the functionality of underground movements, examines the process of recruiting primarily from the viewpoint of the insurgent or revolutionary movement. The companion to this book, the second edition of *Human Factors Considerations of Undergrounds in Insurgencies,* complements this effort by focusing on personal and group motivations and looking at recruitment from the individual's perspective.

PHASES OF AN INSURGENCY

During the early stages of an insurgency, leaders seek to carefully select, investigate, and approach potential fellow insurgents. The theme of early recruiting is the mitigation of security risks. Typically, insurgent leaders establish a revolutionary movement in league with a few close friends who share their ideology and become trusted allies. From these modest beginnings, insurgent leaders will seek out individuals or groups that share their fundamental (if not identical) beliefs.

During the middle stages of an insurgency, leaders usually have to expand the recruiting effort in order to meet growing operational and functional requirements as well as replace members lost to attrition. It is during this transition period that some revolutionary movements fail while others succeed, and the question often comes down to the organization's ability to find new sources for recruitment and support. Successful growth may require an expansion from a rural power base into the cities, or vice versa. It may likewise require a change in the movement's ideology in order for the insurgency to become more inclusive. Or, as has been the case in numerous historical and contemporary insurgencies, the movement's leadership may seek outside membership from foreign fighters and/or mercenaries in order to rapidly acquire the skill sets and expertise required to support the transition.

During the latter stages of an insurgency, recruiting is characterized by the momentum of the movement. A successful insurgency that is able to either take power (supplanting the former government) or achieve political, legal, or quasi-legal status will normally expand recruitment operations to include parallel efforts at political mobilization. Under benign conditions, this transition can feature routine political efforts to attract votes through marketing, propaganda, and political rallies. In the event of a more violent transition into power, mobilization can lead to the formation of local militias that cooperate with the insurgency in the destruction of the regime and its adherents. In this regard, the need to rapidly surge manpower can lead to a requirement to limit the security screening of new members in order to capitalize on an immediate opportunity.

Beyond these general trends, however, each insurgency or resistance movement experiences its own recruiting dynamics, and those dynamics change as the movement matures and moves toward a conclusion. In the case of the Viet Cong, for example, recruiting became more difficult as casualties mounted and villagers became disenchanted with the insurgency. Volunteer selection gave way to forced conscription, and insurgency leaders had to take great care to train and deploy conscripts to keep them isolated from their homes and often from each other to discourage desertion.[2]

RECRUITING TO TASK

Insurgencies must accomplish a wide variety of tasks to be successful, and these activities fall along a spectrum from nonviolent administrative record-keeping to propaganda operations to training and equipping guerilla forces to executing a suicide bombing. Obviously these diverse tasks require different skill sets, different levels of commitment, and different physical and mental capacities. Hence, recruiting operations vary according to the tasks required.

The sensitive nature and long-term investment related to certain tasks require recruitment aimed above all at maintaining security. For example, an insurgent movement that is seeking recruits to develop a systematic intelligence network will carefully vet selectees over an extended period. Conversely, an insurgency that is trying to rapidly fill up the ranks of a conventional or guerilla army in anticipation of stepped-up military operations may use techniques aimed at mass mobilization, with less concern for the political reliability or security risks associated with individuals. As a general trend, therefore, the early phases of an insurgency are marked by very selective screening and evaluation of recruits while the latter stages may evolve into a much less selective, general enrollment approach.

Recruiting Leaders

Leaders tend to join insurgencies and terrorist cells primarily because of their ideological (political/religious/economic/nationalist) views, whereas the rank-and-file members join because of a wide variety of motivations, only some of which touch upon ideology.[3] Hence, insurgent groups looking to recruit and develop senior leaders cull through prospects very selectively. Often, senior leaders are already close associates or friends with current insurgent leaders.

In many cases, senior leaders are not recruited at all, but rather assimilated. The case of Syrian-born jihadist Abu Mus'ab al-Suri demonstrates this trend. al-Suri became radicalized as a young university student and was active both in training and in writing influential jihadist literature long before he became associated with Al Qaeda. He became a confederate of Osama bin Laden after his arrival in Afghanistan and served the organization well by, among other contributions, organizing bin Laden's first taped interviews for the global media. Al-Suri's independent path to jihad, however, also resulted in a sense of detachment from any organization, including Al Qaeda, which he viewed as heading down a wrong path. His biography demonstrates the life of a senior insurgent leader who alternately supported and later denigrated the insurgency he helped to lead.[4]

As an insurgency expands, the underground organization must recruit and develop middle-level leaders, such as front or provincial leaders, influential agents within a university or government agency, and military leaders. Recruiting key figures who already wield great influence in a society is a boon for the insurgent underground because it subtracts the influence and reach of those individuals from the government and adds significant new capability to the movement. A provincial leader, clan elder, or local religious scholar who elects to join an insurgency can potentially provide thousands of recruits from among his constituents.

Closely associated with leader recruitment is the practice of enticing, subverting, or cajoling religious leaders or ideologues into supporting the movement. Popular figures whose ideology or faith can be made to be congruent with the goals of the insurgency can serve indirectly as recruiting tools. The association can be one of cynical convenience—a former KGB agent called such people "useful idiots"[5]—or they can be marriages of true conviction. Lebanese Hizbollah, for example, has from its founding benefited from Shiite clerics who lend religious authority and respectability to the insurgent movement. In this case, the insurgent leaders share the religious faith of the clerics, even if the two groups often differ on matters of strategy and tactics.

Recruiting Soldiers

Insurgencies invariably have a need for security forces, local militias, guerilla armies, and/or conventional military forces to accomplish their objectives. Recruiting fighters thus becomes a major function of insurgent undergrounds.

In the case of the Viet Cong, it was relatively easy to find potential recruits who were inimical to Japanese, French, and later, American invaders. In the wake of World War II, Vietnamese nationalists cultivated a strong sense of resistance toward foreign powers seeking to occupy and exploit their homeland. Because the initial direction of the Communist insurgency was the disruption or nullification of the governmental authority of the South Vietnamese government, it was relatively easy to recruit rural populations and help them organize localized militias. Local militia and guerilla leaders based in villages were the main source of recruiting for the Viet Cong. Later, after 1964, as the Viet Cong began to adjust their aims toward the defeat of American forces, they had to provide manpower for more conventional forces that would fight wherever they were needed. Recruitment therefore became more difficult, and Communist leaders resorted to forced conscription and methods of deception and coercion to keep drafted soldiers in their assigned military units.[6]

25

People recruited into terrorist groups tend to defy profiling,[7] although key factors and considerations for terrorist recruiting can be found in the second edition of *Human Factors Considerations of Undergrounds in Insurgencies*. Nevertheless, case studies on terrorist recruiting suggest some general trends. Among these trends is the not infrequent use of deception to lure an unwitting member into an unintentional "suicide" attack, to include reported use of mentally handicapped men and women in Iraq by Al Qaeda. Technically, of course, the resulting strike is not a suicide but rather a homicide. Still, a small percentage of incidents reported as suicide attacks have included deception of the operatives.

Terrorists, and particularly suicide operatives, often share a strong sense of victimization. Thus, insurgencies seeking such members will use persuasive propaganda methods in order to clarify or exaggerate the degree to which victimization has occurred. Very often members of the insurgency who later engage in acts of terror or suicide operations transitioned into that role having first engaged in more benign activities. Hence, they are recruited for violence only after they have already aligned themselves with the insurgency in other ways. They may have joined protest marches, given money, joined a militia, or housed other operatives, for example, before being selected for a more violent act. However, in extreme cases such as the 1994 Rwandan genocide, a deliberate and prolonged propaganda campaign promoted the perception of victimization for an entire ethnic group and laid the foundation for mass uprising and willful participation in mass atrocities.

Psychologists studying the phenomenon of terrorist recruitment point to the dialectic of incentives pulling one into dangerous or suicidal activities and those pulling in the opposite direction. The latter include family ties (in cases in which the family is not friendly to the insurgency), jobs, and associations with nonviolent organizations. The former include monetary incentives (for the member and/or his family), the perceived impending success of the insurgency, and even intangibles like the respect of elder leaders and group acceptance.[8] The Liberation Tigers of Tamil Eelam (LTTE, or Tamil Tigers) portrayed suicide operations in a semireligious light and cultivated a wide acceptance of such acts as necessary for the hastening of the day when they would achieve autonomy.[9] Other organizations, such as the New People's Army (NPA) in the Philippines, leveraged peer pressure in youth organizations as a means of gaining new recruits, while young foot soldiers in Sierra Leone and Nigeria were often drawn to (or kept in) the Revolutionary United Front (RUF) and the Movement for the

Emancipation of the Niger Delta (MEND), respectively, because of the availability of drugs and money (blood diamonds, bunkered oil, etc.).[a]

TARGET POPULATIONS

Insurgencies recruit successfully from all socioeconomic groups—poor, middle class, and wealthy; urban and rural; men and women; those with criminal records and those without. The mythical idea that only the disaffected or unemployed join insurgencies was laid to rest long ago. Rather, insurgencies in most cases seem able to draw from across the spectrum of society when they choose to. The ideologies, propaganda tools, and recruiting techniques often change based on the nature of the targeted population. In general, however, the motivation for someone joining an insurgency will range from sense of duty (against brutality and evil or religious obligation), nationalism, hatred, despair (nothing left to lose), desire for vengeance, and desire for personal gain.

Rural and urban recruiting differ from each other, primarily because of the security threat, which tends to be more pronounced in the latter. Insurgencies reaching out to recruit members in an urban society tend to be more deliberate and gradual, and they place greater emphasis on ideology because their targeted population is better educated. Conversely, guerilla movements that seek recruits in rural settings can usually afford to make a more direct and immediate approach to peasant communities, and they emphasize opportunity for food and economic success—examples of such movements include the Fuerzas Armadas Revolucionarias de Colombia (Revolutionary Armed Forces of Colombia, or FARC) in Colombia, the Frente Farabundo Martí para la Liberación Nacional (Farabundo Marti National Liberation Front, or FMLN) in El Salvador, and the MEND in Nigeria.[11]

Some scholars have theorized that many of the insurgencies that thrive beyond their native phases operate in societies that have a "culture of radicalization"—i.e., a societal context in which joining a violent or rebellious movement is not considered aberrant, inherently wrong, or disreputable. An example of this is Saudi Arabian tolerance toward Muslim men who fight in religiously motivated wars abroad, even if their behavior is technically illegal.[12] Such societies can inculcate a heroic imagery related to revolutionary behavior among the younger population, facilitating later radicalization.

---

[a] One survey of members of MEND found that many of the young members were drug users—and some drug dealers—with limited education and no economic resources, so they were completely dependent on their leaders for financial support, food, and shelter.[10]

A culture of radicalization benefits from the government's harsh treatment of those suspected of involvement in illegal movements. Police crackdowns, brutal conditions in jails and prisons, and arbitrary violations of person, property, and privacy contribute to feelings of victimization and injustice. This was the case, as an example, for Afghanistan war veterans returning to Saudi Arabia in 2002. Having returned from what they considered a heroic and religiously significant endeavor, they were subjected to suspicion, arrest, and economic marginalization. Many of them joined Al Qaeda as a result.[13] This Saudi Arabian case is ironic, however, in that the police force initially was too lenient and nonintrusive as jihadist leaders and recruiters sought to radicalize the population.[14] Other examples of strong-handed police and security force crackdowns contributing to the radicalization (religious or otherwise) include the killing of Boko Haram founder Mohammed Yusuf in police custody in Northern Nigeria in 2009; the Serbian killing of popular Ushtia Çlirimtare e Kosovës (Kosovo Liberation Army, or KLA) leader Adem Jashrai and his family in Kosovo in 1998; El Salvadoran "death squads" targeting FMLN sympathizers in the 1980s; and the British Army's use of live ammunition against Irish protesters in Derry in 1972 ("Bloody Sunday").

Another societal factor that contributes to radicalization and recruitment is unemployment or underemployment. As noted above, this condition is not determinative—wealthy, employed people sometimes join insurgencies, and many unemployed/underemployed people do not—but it is often a strong contributing factor. Well-documented cases include the "lumpen" youth who joined the Revolutionary United Front in Sierra Leone and recently laid-off coffee plantation workers in Rwanda who were deliberately drawn into soccer clubs and indoctrinated with the concept of "Hutu Power."

Rural Populations

Insurgencies often have unfettered access to rural populations, chiefly because of their remoteness from centers of government power. Typically, such communities are poor and produce sufficient numbers of unemployed or underemployed young people to attract insurgent groups. Recruiters can offer targeted youngsters the promise of food, money, and upward mobility. Examples of this approach are many. The Colombian insurgency FARC, for example, reached out to poor, uneducated rural youth and emphasized not its Marxist ideology but rather "three square meals per day" and the vision of a prosperous future.[15]

The Viet Cong based much of their recruiting efforts in the rural communities of South Vietnam, in part because the government had little influence there. But village recruiting, particularly in the Central Highlands of Vietnam, was not just the result of opportunity; it was part of the overall objective of limiting and disrupting the reach of the government in Saigon. The creation and sustainment of a rural power base not only sustained the insurgency with a reliable source of manpower, but it also demonstrated the weakness of the American-backed regime.[16]

Urban Populations

Although the remoteness of rural villages attracts insurgent recruitment, a successful revolutionary movement must at some point expand into urban areas where, in most modern societies, the bulk of the population lives. Urban recruiting carries with it increased security risks due to the proximity of government forces, but it also has the advantages of being able to reach a huge population base and providing the recruiters with the opportunity to hide among the masses.

In the 1970s, FARC had limited reach and impact on society, mainly because it remained largely a rural movement. As the country continued to urbanize, however, leaders decided that they would have to expand operations into the cities. The opportunity came as poor urban workers began to protest against their living conditions and economic stagnation. The FARC secretariat quickly attached themselves to this grievance and represented the movement as that of the proletarian struggle against the imperialism and corruption of the government. They began to establish student groups and civic action programs within universities and schools, and they used these platforms to persuade people to vote for left-wing politicians and agitate for reforms that would benefit the insurgency. FARC also drew upon the burgeoning urban population for recruiting into local militias and mobile guerilla armies. Leaders learned to adapt their recruiting methods to various target audiences. Whereas rural recruits were typically drawn into the insurgency with promises of basic necessities, urban youth responded more to strong ideological propaganda.[17]

Elites

Upper-class or political elites often get involved in insurgencies for reasons entirely different from those that draw in their underprivileged fellow citizens. Whereas lower classes are often enticed into insurgent movements with incentives related to the basic necessities of life—food, shelter, protection, money—elites generally join insurgencies because of ideological conviction, sometimes coupled with personal grievances. As highlighted earlier, however, upper-class and political elites may be high-profile opposition members, which would make their membership in an underground a potential security risk for the organization.

Military

Insurgencies often reach out to military personnel for recruitment into the movement, and not just established, conventional forces, but also illegal or quasi-legal guerilla forces. Al Qaeda on the Arabian Peninsula, for example, recruited heavily from the ranks of those Saudis who had fought in Afghanistan in the late 1990s and then returned home. These men offered several advantages to the radical movement: they were trained in combat techniques, they had proven themselves as religiously committed, and they shared a common and strong bond of camaraderie with their fellow veterans. These attributes make for strong and reliable insurgents.[18] A second example is the "sobel" (soldier-rebel) phenomenon in Sierra Leone in which soldiers would join their rebel friends at night to share drugs and alcohol, periodically conduct joint raids on villages to steal property, and jointly run illegal diamond-mining operations.[19]

Professional Soldiers/Private Military Contractors

Although not technically "recruited" into an insurgency, professional soldiers provide an immediate and critical force-multiplier effect for ill-trained and poorly armed insurgent forces, and their impact on an insurgency cannot be dismissed. Despite having their loyalty to a movement defined by their contracts, these warfighting experts can fill critical functional roles in an insurgency, such as training and logistics, or in some cases as security forces or actual foot soldiers.[b] The ability to suddenly introduce a Hind gunship where no airborne asset had previously existed or to develop and deploy advance sniper team capabilities are but two examples of recent insurgency operations supported by professional soldiers in the Balkans and West Africa. Conversely, it should be noted that governments under siege by insurgent forces have also resorted to the use of professional soldiers, which has led to a range of interesting scenarios such as FARC insurgents serving as hired snipers to defend Colonel Muammar Gaddafi in November 2011 during the Libyan uprising and mercenaries who served together in Angola

---

[b] It is not our intent in this section to argue the differences between private military contractors (PMCs), professional soldiers, and mercenaries. Rather, the point of this discussion is to underscore the fact that insurgent forces have in the past used and will continue in the future to use these expert resources when there is a need and the insurgency has the financial means to do so.

finding themselves supporting two different contracts and opposing forces a year later in Sierra Leone.[20]

Women

Modern insurgencies often recruit women for a variety of reasons. The most common one is the need for numbers to fill the ranks of military forces, especially when operations deplete the ranks of men. Female insurgents also prove useful in terrorist and suicide operations, particularly when cultural norms prevent or discourage close monitoring of or searching for women. Marxist and other modern movements sometimes use women recruits for propaganda purposes to demonstrate their modernism and friendliness to women's rights. The LTTE targeted women for recruitment as a means for siphoning off strength from other Tamil resistance movements.[21] The ruthlessness of Eritrean female fighters during the insurgency against Ethiopia became legendary.[22] FARC brought women into the movement and encouraged romantic relationships in order to strengthen recruitment and sustainment. Pregnancies, however, were discouraged, and pregnant females were directed to either leave the movement or have an abortion.[23] For women, recruitment carries with it not only the danger of arrest, wounding, or death, but also the prospect of liberation from traditionally patriarchal societal norms. Thus, the practice can lead to a transformation of society in favor of women if the insurgency is successful.

Children

Recruitment of children remains a politically volatile practice, but modern insurgencies turn toward this option as a way of swelling the ranks, replacing losses, and obtaining operatives that can evade detection. This category includes examples of recruiting and training teenagers for fighting and eventual integration into existing units or at times forming special units composed of youngsters.[24] But it also includes the use of younger children, particularly in urban areas, for the purposes of intelligence, courier duties, or even terrorism. For purposes of this chapter, children include youths from approximately age five through twenty-two—i.e., from kindergarten through college.

Over half of the groups designated as terrorist organizations by the U.S. government employ children in their movements. These groups include Al Shabaab, Ḥarakat al-Muqāwamah al-'Islāmiyyah (Islamic Resistance Movement, or HAMAS), Hizbollah, Al Qaeda, Jemaah Islamiyah, and Euskadi Ta Askatasuna (Basque Homeland and Freedom, or ETA). Inducements include money, camaraderie, and excitement. Most groups seeking youngsters extend their efforts through

31

local schools and use the Internet to reach a global audience. In some cases, youth are recruited by force or are duped into participation, including participation in suicide attacks.[25]

Within countries and regions that are torn with violence and lawlessness, young people are born into societies predisposed to radicalization. The absence of social mobility, meaningful employment, public safety, and hope for a stable and prosperous future makes the metamorphosis into a terrorist or insurgent seem normal if not commendable. Conversely, in societies in which there is the rule of law, minors can engage in illegal activities, including violent movements, with some degree of impunity because of their age and legal status.

FARC's use of children numbered in the thousands, and they sometimes used children to bolster their recruitment efforts, particularly in urban environments. The deliberate use of attractive and popular young people often led to mass recruitment among the young. Although condemned by human rights groups, the practice of recruiting youth was represented by FARC officials as being a necessary effort to care for youngsters whose parents had been unable to provide for them.[26]

As with other targeted populations, insurgencies recruit from a wide variety of children—children from the lower, middle, and upper classes; children with various levels of education and intelligence; and children who fall along a wide spectrum of ideological fervor. Likewise, insurgent recruiters are likely to employ "precursor" groups and activities—e.g., student groups, athletic clubs, religious organizations, refugee camps, etc.—from which to evaluate and draw young people into the movement. Religious leaders and teachers, along with family members and peers, are most often involved in the actual recruiting of youth, although the Internet is also effective at encouraging young people to "self-radicalize."[27] In addition, insurgencies often view regions with large displaced populations as prime breeding grounds for the recruitment of youth because children from those populations are vulnerable, accessible, and often unattached to family or other protective groups.

Of the methods used to recruit youth, four are most common. Some children are born into an environment that encourages radicalization, such as war-torn Afghanistan or Somalia. In these environments, insurgent organizations can fairly easily round up children, offering them food, shelter, protection, and acceptance. In some cases, youngsters are forced into an insurgency, often through kidnapping. Others are persuaded to join a cause. Finally, there are those who become "self-radicalized."[28]

TECHNIQUES

One of the consistently observed trends in recruiting techniques of modern insurgencies is that of inducing people to join benign and easily accessible groups as a precursor to later recruitment into the insurgency. People who join student groups, protest marches, or other ideology-based movements can often do so at little risk because the activities are nonviolent, legal, or at any rate low risk. Such groups then serve as useful recruiting pools because they comprise large numbers of people who are already somewhat aligned with insurgent ideology and who are less likely to be government agents.[29]

It is not surprising, then, that actual recruitment of individuals into an insurgency takes an extended period of time, even years. The recruitment process gradually brings together a developing individual and a careful, methodical insurgent group concerned about security. When operatives judge that the individual possesses the requisite enthusiasm and abilities and is an acceptable security risk, they approach the prospective member. Some insurgents report that it took a while for them to realize they were being recruited. The recruit might spend a lengthy period in an entry-level status so that the leadership can assess his or her reliability and potential. Later, the recruit may advance in the organization into increasingly committed levels of activity. One insurgent described the levels of advancement as sympathizer, collaborator, premilitant, and militant.[30]

An examination of recruiting techniques reveals a system in which the leadership of an insurgency is actively seeking out people to recruit and then acting on the targeted persons to induce them to join. While this is sometimes the case, it by no means characterizes every insurgency. For self-radicalized Muslims, the norm appears to be for them to seek out opportunities to join a violent movement, rather than the reverse.[31]

Ideological Appeal/Radicalization

Ideology plays a role in every insurgency and revolutionary movement. The *raison d'être* of the movement may be relatively simple, like revenge against repressive treatment or racial bigotry. More often in modern insurgencies the ideology draws from established religious, political, or social principles. The leadership of an insurgency must integrate the ideology with a salable body of propaganda that appeals to targeted audiences, and usually this message deals with the perceived injustice of the status quo. For example, Islamic insurgencies look to religious ideology but then contrast these ideals with the current situation of the population. The insurgency then represents itself

33

as a legitimate attempt to right the wrongs of society. Because the status quo is not in line with the ideals of Islam, the insurgents can pose (indeed, even think of themselves) as agents of God.

An example of the centrality of ideology to the process of recruitment is the large body of Saudi Arabian men who were drawn to the fighting in Afghanistan in the late 1990s and then later recruited into Al Qaeda. Most of these young men were enticed into activism through propaganda—video, Internet, booklets, and sermons—that highlighted the horrors of the Chechen war and the tribulations faced by the Taliban regime. Recruiters portrayed these conditions as evil attacks against Islam and encouraged their audiences to do their religious duty by traveling to Afghanistan to fight. Their message was powerful, persuasive, and successful, largely because it drew upon strong religious themes.[32]

The role of religion in both insurgency and counterinsurgency is a compelling variable, depending on the historical context. In conflicts throughout history, opponents were often co-religionists, praying to the same God to strengthen their hands against the other. In such cases, the tendency was to view the adversary as one who opposed God's will. In other cases, prominent and powerful religious organizations allied with the government, causing those in the insurgent movement to either embrace atheism or reject the religious authority of the establishment. That said, it remains true that in many cases of insurgency, religious motivation plays no role whatsoever, while in others it can be a primary component of rebellion against the authority of the government.

In Islamic insurgencies and terrorist groups, religious affiliation is a *sine qua non* to joining. Past studies have shown that typical members of Islamic insurgencies are Muslims who have greater than average religious convictions. They also tend to be somewhat educated, middle class, married with children, and underemployed.[33] Religious conviction alone, however, is not sufficient to induce recruitment. The other key factor is personal ties to family, friends, and other associates.

In the case of Pakistani jihadists, religious conviction became a factor as part of the process of radicalization, rather than at the start. One study examined the process of jihadist leaders recruiting young men and teenagers into camps, where members found themselves forming close personal bonds—eating, working, sleeping, and studying together with other disaffected youths. Religious training and exposure to propaganda highlighting, for example, the rape of Muslim women in Kashmir and Bosnia became a part of their daily regime. This process, in turn, produced strong feelings within the youngsters that they had a religious duty to fight in order to redress these wrongs against God and their fellow Muslims.[34]

Socioeconomic and political ideology functions in a similar way by pointing to the status quo and demonstrating that the government's failed policies, corruption, and repression have caused undue misery— poverty, disenfranchisement, famine, disease, etc.—among the people. The insurgents, by way of contrast, embrace and represent the correct and responsible path to success. The intersection of criminal gangs and insurgent groups in the Niger Delta has created a pseudo-Robin Hood persona in which the "bunkering" or stealing of oil from pipelines and ships is viewed as a return of the region's natural riches from the Nigerian federal government and their multinational oil partners who are stealing the resources. This is despite the fact that many of these criminal gangs keep their ill-gotten oil revenues for their personal use.[35]

Racial and nationalist movements develop ideologies based on ethnic differences, but they tie these differences to the resulting status quo. The most obvious example of this is resistance to invasion or occupation by a foreign or imperialistic power. The Irish Republican Army, for example, viewed the British government as an unjust foreign ruler and linked its presence with the unjust status quo. The Tamil Tigers likewise fell into this category, but their goal was to carve out an autonomous state within Sri Lanka.[36] In Burma, the Karen National Union/Karen National Liberation Army deliberately fostered an ethno-nationalist culture and promoted this culture at an early age in schools through the teaching of "ethno-history" as well as the adoptions of nationalist symbols, such as a coat of arms, national dress, and a national anthem.[37] In these examples, the insurgents viewed themselves as representing a separate and distinct people—a nation distinguished by race, religion, culture, and/or language—against a dominant, repressive regime.

Finally, it is instructive to view the role of catalyzing events and their relationship to ideology in the recruitment of insurgents and revolutionaries. Ideological principles and iconography can underlie a prospective insurgent's worldview for years and yet not stimulate him or her to activism. But when dramatic events occur—either on the world stage or in one's personal life—these events are often colored by one's beliefs and can stir the person toward a violent reaction. This phenomenon was illustrated by the publication of derogatory cartoons of the Prophet Muhammad in Denmark in 2005. In the wake of the publication, many Muslim jihadists demanded a vigorous response from Muslims worldwide and lamented what they viewed as the pathetic weakness of the umma (i.e., the global community of Muslims) in failing to respond. Others, however, saw some good in the incident, because the outrage served to awaken Muslims who otherwise might have remained dormant. Hence, the perceived attack on Islam served as a recruiting tool and a call to action.[38]

Insurgencies, especially during later phases of the movement, sometimes offer incentives to potential recruits. Typically these incentives include jobs, food, and/or money. Less often they include protection from criminal arrest and prosecution, needed medical services, and even sex. Intangible incentives are equally important: group acceptance, the respect of elders, and the sense of heroic self-actualization.

Other forms of incentives include the provision of a refuge for the disaffected. Some members drift into an insurgency and join it largely because of personal problems or issues unrelated to the group's ideology.[39] Others join as youths looking for an opportunity to rebel or have an adventure.[40] Child soldiers fighting with the RUF in Sierra Leone were enticed by a range of these incentives—they received food, drugs, and money from their RUF leaders and received the status and companionship that came from being part of the militant organization.[41]

Coercion/Conscription/Deception

Insurgent groups that achieve and maintain control over a region are able to include coercion or conscription in their recruitment efforts. The LTTE, for example, reportedly conscripted one son from each family living in the regions they controlled during the 1990s. Such methods were aided by the climate of resentment toward the Sri Lanka army's repressive and violent acts against the Tamil people.[42] Bribery or blackmail can also be employed as effective tools for recruiting personnel with key functional skills that are needed in the insurgency, such as bankers and financial experts, chemists and other technical experts, or government bureaucrats with access to critical information sources.

Personal Ties

The development or exploitation of personal ties to potential recruits is a common feature in insurgency recruiting operations. Relations between recruiters and targets can emanate from family ties, religious affiliation, student groups, and other activities. In societies in which there is a "revolutionary climate," existing personal ties are a key motivator for someone to join an insurgency. This type of environment exists in areas in which there has been prolonged conflict that has touched a large majority of the population. For example, the population of Colombia experienced some 300,000 deaths due to political violence between liberals and conservatives prior to the emergence of insurgent groups in the 1960s.[43] In other cases, recruiters develop relations with potential members in order to vet them and mitigate security risks.

The role of families is crucial both to recruiting and to efforts aimed at combating it. In societies in which the family plays a prominent role in the lives of young adults, the household setting and family history can in fact become the most determinative factor. Families that have a history of disaffection from the government or actual violence against it can create an atmosphere that encourages young people to join the fight. Conversely, families that oppose the insurgency exert a strong normative influence on young people that makes them resistant. In either case, recruits sometimes break ties with their family either because of the family's opposition to the insurgency or in order to protect the family from government suspicion and violence.

Prison Recruiting

Prison recruiting takes advantage of a literally captive audience to further swell the ranks of insurgencies and terrorist organizations. The Korean War saw one of history's most extensive infiltrations of a prison population—the Communists' takeover of the prisoner-of-war camp at Koje-do. In that case, the insurgents used suasion, coercion, and violence to subvert large sectors of the prison population and eventually set up courts, executed those convicted of crimes against the Communists, warred against anti-Communist factions, and triggered armed intervention by the U.S. Army. The incident was used as propaganda against the West and remains a stunning example of how prisons can become breeding grounds for rebellion and violence.[44]

Today the threat resides in correctional facilities in which prisoners are vulnerable to faith-based recruitment. The typical pattern features criminals who are already radicalized entering the prison system and subsequently converting other inmates to religious beliefs that lead or can lead to radicalization and membership in an insurgency or terrorist cell. The threat becomes active upon the convert's release from prison. Thus, the threat most often emanates from within the prison population rather than from outside sources.

Convicted criminals residing in prison often look to religious beliefs for solace when faced with serving long sentences in the difficult, threatening, and unnatural setting of prison life. Religious faith can be a positive influence in the lives of prisoners, but in some cases it can also be a pathway to recruitment into a violent movement. Some prisons report the merging of religious influences and gang dynamics, resulting in conversions aimed at filling the ranks of various gangs or co-opting gang organizations into radical religious movements.

Individual prisoners often turn to radical religions partly because of genuine spiritual convictions and partly for protection.[45]

The actual threat of prison recruitment into active terrorist cells in the West remains small but real. In 1997 an American convert to Islam, Kevin James, founded a movement he called Jam'iyyat Ul-Islam Is-Saheeh. In 2004 he converted a fellow inmate at New Folsom Prison, Levar Washington, who, upon his release, converted and recruited two more men into the terrorist cell. They planned to conduct bombings of various targets in the Los Angeles area and began by robbing several gas stations at gunpoint. They were arrested before the plots could develop.[46]

Radical groups reportedly consider prison populations as fertile grounds for recruitment. Maximum-security facilities remain the best sites for recruitment because they are often overcrowded, lack institutional resources that could provide inmates with moderating spiritual guidance, and have serious problems with gangs. The potential for faith-based or ideology-based conversion and radicalization in such conditions is high, and the U.S. Department of Justice has instituted studies of the problem with a view toward mitigating the risks.

Subversion

Subversion is defined as being "designed to undermine the military, economic, psychological, or political strength or morale of a governing authority."[47] As it relates to insurgent recruiting, it also carries the meaning of corrupting someone's morals or loyalties. Whereas other methods of recruiting can seek to win over the general population, subversion takes aim at those within the government—i.e., officials, administrators, government workers, police, or military personnel.

One of the best examples of the use of subversion as a recruiting technique was the Viet Cong doctrine of *binh van*—i.e., the promotion of desertion and defection from the government of Vietnam. Through agitation, persuasion, coercion, and threats, Viet Cong operatives targeted key officials, both military and civilian, to weaken the government's ability to rule as well as to swell the ranks of the insurgency. Subversion against the military was effective in diminishing the soldier's will to fight and would in some cases succeed in causing him to provide intelligence on military operations or even to defect to the Viet Cong.[48]

Undergrounds seek to infiltrate communication and transportation industries, often through subversion, because agents therein can sabotage facilities needed for the mobilization of the military and police forces and can provide intelligence on mobilization schedules. Unions are also prime targets because the control of these groups enables an underground to call strikes, weaken governmental control, or cause general social disorganization. Subversion of labor unions is also desirable because union funds can be diverted to underground activities. Underground funds may also be concealed in union accounts by falsifying the records. Strikes, demonstrations, and riots also diminish the effectiveness of the government forces. Police, militia, and regular army troops may be required to control them, and this draws manpower from the units assigned to combating the underground. Punitive measures taken by the police and injuries suffered by participants or onlookers are exploited by underground agitators to turn minor skirmishes into major incidents. By exploiting the resulting agitation, an underground may be able to rally the people to the revolutionary movement and disrupt government control. The classic example of the power of labor unions is the Solidarity movement in Poland during the 1980s and the eventual transformation of that labor movement and insurgency into a political party that undermined the authority of the Communist Party and placed Lech Walesa as president of Poland in 1990.[49]

CONCLUSION

Analysis of an underground's recruitment operations requires understanding of the movement's ideology, the security environment in which it operates, the phasing of the insurgency, the required skills being recruited, the target populations, and the consequent techniques employed by leaders and their agents. Because recruiting operations are conditioned by these contexts, they tend to evolve with the progress (or lack of progress) of the insurgent movement. The overall success or failure of recruitment is a key indicator of the success or failure of the insurgency or resistance itself, because without a reliable supply of leaders, workers, and soldiers, the underground cannot sustain resistance.

ENDNOTES

[1]   Shana Marshall, "Hizbullah: 1982–2009," in *Casebook on Insurgency and Revolutionary Warfare, Volume II: 1962–2009*, ed. Chuck Crossett (Laurel, MD: The Johns Hopkins University Applied Physics Laboratory, 2010), 379–380, 387–388.

[2]   Bryan Gervais, "Viet Cong 1954–1976," in *Casebook on Insurgency and Revolutionary Warfare, Volume II: 1962–2009*, ed. Chuck Crossett (Laurel, MD: The Johns Hopkins University Applied Physics Laboratory, 2010), 335–336.

[3]   Petter Nesser, "Jihad in Europe: Recruitment for Terror Cells in Europe," in *Paths to Global Jihad*, eds. Laila Bokhari, Thomas Hegghammer, Brynjar Lia, Petter Nesser, and Truls H. Tønnessen (Kjeller, Norway: Norwegian Defence Research Establishment, 2006), 10.

[4] Brynjar Lia, "The Al-Qaida Strategist Abu Mus'ab al-Suri: A Profile," in *Paths to Global Jihad*, eds. Laila Bokhari, Thomas Hegghammer, Brynjar Lia, Petter Nesser, and Truls H. Tønnessen (Kjeller, Norway: Norwegian Defence Research Establishment, 2006), 39–53.

[5] 1985 interview with Yuri Bezmenov, posted by MHadden88, "Bezmenov on Marxists," *YouTube*, uploaded October 5, 2008, http://www.youtube.com/watch?gl=US&hl=uk&v=dE38dLxapVo.

[6] Bryan Gervais and Jerome Conley, "Viet Cong: National Liberation Front for South Vietnam," in *Assessing Revolutionary and Insurgent Strategies*, ed. Chuck Crossett (Laurel, MD: The Johns Hopkins University Applied Physics Laboratory, 2009), 36–38.

[7] John Horgan, "From Profiles to Pathways: The Road to Recruitment," *Foreign Policy Agenda* 12, no. 5 (2007): 24–27.

[8] Ibid.

[9] Michael J. Deane and Maegen Nix, "Liberation Tigers of Tamil Eelam (LTTE)," in *Assessing Revolutionary and Insurgent Strategies*, ed. Chuck Crossett (Laurel, MD: The Johns Hopkins University Applied Physics Laboratory, 2009), 49.

[10] Judith Burdin Asuni, "Understanding the Armed Groups of the Niger Delta," (working paper, Council on Foreign Relations, New York, September 2009), 7.

[11] Mauricio Florez-Morris, "Joining Guerilla Groups in Colombia: Individual Motivations and Processes for Entering a Violent Organization," *Studies in Conflict and Terrorism* 30, no. 7 (2007): 626.

[12] Thomas Hegghammer, "Militant Islam in Saudi Arabia: Patterns of Recruitment to 'Al-Qaida on the Arabian Peninsula,'" in *Paths to Global Jihad*, eds. Laila Bokhari, Thomas Hegghammer, Brynjar Lia, Petter Nesser, and Truls H. Tønnessen (Kjeller, Norway: Norwegian Defence Research Establishment, 2006), 23, 28–29.

[13] Ibid., 27.

[14] Ibid., 28–29.

[15] Stephen Phillips, "Fuerzas Armadas Revolucionarias de Colombia—FARC," in *Assessing Revolutionary and Insurgent Strategies*, ed. Chuck Crossett (Laurel, MD: The Johns Hopkins University Applied Physics Laboratory, 2010), 26.

[16] Gervais and Conley, "Viet Cong," 335–336.

[17] Phillips, "FARC," 21–22. An exception would be urban youth in some underdeveloped countries who experienced very high unemployment and had basic needs just as their rural brethren.

[18] Hegghammer, "Militant Islam," 25.

[19] Jerome Conley, "The Revolutionary United Front (RUF), Sierra Leone," in *Casebook on Insurgency and Revolutionary Warfare, Volume II: 1962–2009*, ed. Chuck Crossett (Laurel, MD: The Johns Hopkins University Applied Physics Laboratory, 2009).

[20] Robin Yapp and Sao Paulo, "Female Colombian Snipers 'Fighting to Defend Col Gaddafi in Libya,'" *The Telegraph*, April 14, 2011, http://www.telegraph.co.uk/news/worldnews/africaandindianocean/libya/8451467/Female-Colombian-snipers-fighting-to-defend-Col-Gaddafi-in-Libya.html; author's interview with personnel involved in the Sierra Leone and Angola operations, South Africa, August 2010.

[21] Deane and Nix, "Liberation Tigers," 48.

[22] Jerome Conley, interview with former rebel commanders, Asmara, Eritrea, June 1997.

[23] Phillips, "FARC," 27.

[24] Deane and Nix, "Liberation Tigers," 48.

[25] Catherine Bott, W. James Castan, Rosemary Lark, and George Thompson, *Recruitment and Radicalization of School-Aged Youth by International Terrorist Groups*, Final Report (Arlington, VA: Homeland Security Institute, 2009), 1.

[26] Phillips, "FARC," 26–27.

[27] Phillips, "FARC," 12–13; Bott et al., *Recruitment and Radicalization*, 1.

[28] Bott et al., *Recruitment and Radicalization*, 14.

[29] Florez-Morris, "Joining Guerilla Groups," 620–621, 625.

[30] Ibid.

[31] Nesser, "Jihad in Europe," 11.

[32] Hegghammer, "Militant Islam," 26–27.

[33] Nesser, "Jihad in Europe," 10.

[34] Laila Bokhari, "Paths to Jihad—Faces of Terrorism: Interviews Within Radical Islamist Movements in Pakistan," in *Paths to Global Jihad*, eds. Laila Bokhari, Thomas Hegghammer, Brynjar Lia, Petter Nesser, and Truls H. Tønnessen (Kjeller, Norway: Norwegian Defence Research Establishment, 2006), 35–36.

[35] Jerome Conley, "The Movement for the Emancipation of the Niger Delta (MEND)," in *Casebook on Insurgency and Revolutionary Warfare, Volume II: 1962–2009*, ed. Chuck Crossett (Laurel, MD: The Johns Hopkins University Applied Physics Laboratory, 2009).

[36] Deane and Nix, "Liberation Tigers," 41–42.

[37] Ron Buikema and Matt Burger, "Karen National Liberation Army (KNLA), Burma," in *Casebook on Insurgency and Revolutionary Warfare, Volume II: 1962–2009*, ed. Chuck Crossett (Laurel, MD: The Johns Hopkins University Applied Physics Laboratory, 2009).

[38] Truls H. Tonnessen, "Jihadist Reaction to the Mohammed Cartoons," in *Paths to Global Jihad*, eds. Laila Bokhari, Thomas Hegghammer, Brynjar Lia, Petter Nesser, and Truls H. Tønnessen (Kjeller, Norway: Norwegian Defence Research Establishment, 2006), 64.

[39] Nesser, "Jihad in Europe," 11–12.

[40] Ibid., 20.

[41] Conley, "RUF."

[42] Deane and Nix, "Liberation Tigers," 47.

[43] Florez-Morris, "Joining Guerilla Groups," 616.

[44] There are several good histories of the events at Koje-do. See, for example, Robert O'Brien, *Barbed Wire Battleground* (Victoria, BC: Trafford Publishing, 2006); and T. R. Fehrenbach, *This Kind of War* (Washington, DC: Brassey's, 1963).

[45] Mark S. Hamm, *Terrorist Recruitment in American Correctional Institutions: An Exploratory Study of Non-Traditional Faith Groups* (Washington, DC: National Institute of Justice, 2007), 4–6.

[46] "Man Involved in Domestic Terrorism Plot Targeting Military and Jewish Facilities Sentenced to 22 Years," Department of Justice Memorandum, June 23, 2008, http://www.justice.gov/opa/pr/2008/June/08-nsd-556.html.

[47] Joint Publication 3-24, *Counterinsurgency Operations* (Washington, DC: Department of Defense, 2009), GL-9.

[48] James I. Wirtz, *The Tet Offensive: Intelligence Failure in War* (New York: Cornell University Press, 1991), 22.

[49] Chuck Crossett and Summer Newton, "Solidarity," in *Casebook on Insurgency and Revolutionary Warfare, Volume II: 1962–2009*, ed. Chuck Crossett (Laurel, MD: The Johns Hopkins University Applied Physics Laboratory, 2009).

CHAPTER 3.

INTELLIGENCE

CHAPTER CONTENTS

Robert Leonhard and Jerome M. Conley

INTRODUCTION

One of the key functions of an underground—whether an insurgency or a resistance against a foreign invader—is intelligence. Leaders must have relevant and timely information in order to make decisions and lead their organizations. Hence, underground operations typically include provisions for the systematized collection of intelligence on enemy forces (both police and military) and dispositions, political developments, lucrative targets for sabotage or guerilla action, defectors, population dynamics, criminal and law enforcement activities, and a variety of other factors. Intelligence thus feeds senior-level decision making as well as small-unit tactics and virtually every other activity of the underground, the guerilla force, and the public component.

In this endeavor, the insurgency or resistance usually has the advantage of knowing the terrain and the people. Drawing on the strength of clandestine networks and friendships, undergrounds often have access to intelligence by virtue of having confederates operating throughout the theater of conflict. The Viet Minh (later, Viet Cong) insurgency demonstrated this strength. Pursuant to the 1954 Geneva Accords, large-scale regrouping of the population both north and south of the 17th parallel left underground operatives distributed throughout the country, connected with each other through clandestine networks. The insurgents in the south were perfectly positioned to provide military and political intelligence to the Communist leadership in Hanoi.[1]

The Communist New People's Army (NPA), which grew to its zenith during the presidency of Ferdinand Marcos (1965–1986), likewise demonstrated the efficacy of cultivating close relations with rural communities. By providing impoverished and disenfranchised peasants with real economic advocacy and opportunities for democratic organization, the NPA was able to win the loyalty of the people, who in turn provided accurate, timely intelligence concerning government operations and countermeasures.[2]

In the case of the Provisional Irish Republican Army (PIRA), Republican insurgents were successful in preventing British security forces from infiltrating base areas and gaining intelligence by establishing close relations with their communities and using both coercion and incentives to encourage loyalty. Eventually government forces intensified their efforts to gather intelligence through informers, surveillance, and interrogation.[3] The struggle for control of intelligence continued throughout the 1970s, and the increasing success of British efforts led to the PIRA changing its organization and practices to better secure itself. PIRA members began to organize in small cells rather than in the larger battalions of the earlier years. They also took greater care in instructing recruits to refrain from discussing operations with anyone. Their sophisticated response to British intelligence efforts included training members to avoid leaving forensic evidence after an operation and how to resist interrogation after capture.

Underground intelligence networks most often extend beyond the borders of the movement's native country. It is common for undergrounds to have cells distributed throughout the world among populations sympathetic to the cause. The Liberation Tigers of Tamil Eelam (LTTE), for example, maintained more than fifty offices and cells in foreign countries, especially in countries with large numbers of Tamil expatriates, such as England, France, Australia, Canada, and the United States. One Canadian intelligence report noted that the LTTE had communication hubs in Singapore and Hong Kong to facilitate its weapons procurement activities, with secondary cells in Thailand, Pakistan, and Myanmar and front companies in Europe and Africa. From these locales, LTTE operatives coordinated purchases and shipments from Asia, Eastern Europe, the Middle East, and Africa.[4]

This chapter examines specific intelligence practices of insurgent and resistance movements from World War II through the present.

MILITARY INTELLIGENCE

Insurgent and revolutionary movements typically operate from a position of numerical and technological inferiority vis-à-vis government forces. Offensive operations therefore require accurate intelligence in order to maximize effectiveness and maintain security of often small, vulnerable guerilla forces. Undergrounds provide this critically important military intelligence not through normal military methods but rather through their infiltration of society and their networks of clandestine operatives.

Underground workers (and their associated operatives within the auxiliary) assist guerrilla forces by providing valuable data about the enemy and the area of impending combat. This information may include the number of enemy troops, their deployment, their unit designations, the nature of their arms and equipment, the location of their supply depots, the placement of their minefields, the pattern and routine of their patrols, and their morale, as well as various topographical factors, such as swamps and ravines that govern access to enemy emplacements. Sometimes this information is obtained directly by underground personnel through visual observation of the targets.

46

For example, members of the French resistance reconnoitered German coastal defenses in the preparation for the Allied invasion of France in June 1944. During the early years of World War II, Vietnamese resistance leader Ho Chi Minh and the Viet Minh worked with the American Office of Strategic Services against Japanese forces, including providing intelligence on Japanese dispositions and activities. Such data may also be collected by auxiliaries in the local populace, or "popular antennae," as these sources are described in one Viet Minh manual. The Viet Minh used children playing near Japanese (and later French) fortifications as a source of information on troop arrivals and departures, the guard system, and other pertinent details—all of which were easily observable by untrained children—that aided the guerrillas in planning attacks.

Undergrounds often excel at infiltrating government agencies and facilities, with the result that they have access to crucial military intelligence that can prove decisive. In March 2007, the LTTE launched an air attack on the Sri Lankan Army's Katunayake air force base. The attack was successful mainly because the base's radar systems were undergoing maintenance and were therefore offline. The attackers obviously timed the attack based on this piece of critical intelligence, indicating that the underground had infiltrated the base or its command system. The LTTE's intelligence training came from the Research and Analysis Wing in India during the 1970s when Tamil separatist groups were trained in camps in Tamil Nadu. Early in the movement, Thiruvenkadam Velupillai Prabhakaran, senior leader of the movement, instructed operatives to collect the intelligence manuals of other countries—including the United States, United Kingdom, and Israel—and to translate them into Tamil.[5]

Most PIRA members kept day jobs and took part in operations during the weekends or evenings. Their regular jobs often helped in some way to support the PIRA. If they worked in a government office, gaining official paperwork/forms or intelligence was part of their duty. Catholics in government administrative jobs provided rich information, such as the home addresses of policemen or loyalist paramilitary members, to the Provisionals.[6]

Resistance movements that are allied with external powers or an exiled government thus provide invaluable intelligence aimed at the defeat of the occupying power. Intelligence activities in this context are generally conducted under the guidance of outside governments or companion military forces in the field. These sponsors not only assign targets for reconnaissance but also give technical direction, because most underground personnel lack experience in this type of work. For example, in World War II, specially trained "Jedburgh" teams (composed of an American, a Briton, and a Frenchman) were sent into France to guide resistance workers in their intelligence surveys. These teams were also equipped to conduct the necessary radio communications with Britain. Likewise, Red Army personnel were assigned to the Soviet partisans to direct these activities. When military personnel have not been available to give instruction, underground members have been instructed by manuals. This was done during World War II in the Soviet Union where detailed booklets such as the *Guide Book for Partisans* were circulated for use in regions under German occupation. The following excerpt from a passage in this manual is a typical instruction:

> If you happen to encounter troops . . . do not show that you observe the enemy . . . ascertain the colour of their headgear, their collar braid, and the figures on their shoulder straps. If they have questioned the inhabitants about something, try to find out what the Fascists have asked . . .[7]

This manual also gives tips for ascertaining enemy intentions: if an attack is planned, trucks will arrive loaded and depart empty; if the enemy intends to retreat, fuel and foodstuffs will be removed, roads and bridges will be demolished, telephone wires will be removed, and trains and trucks will arrive empty and depart full.

Undergrounds cooperate with guerilla forces in the intelligence function, and their activities most often overlap and complement each other. In South Vietnam, for example, Viet Cong underground operatives in local villages would often provide military intelligence to local militia groups, who would in turn add to the intelligence picture in support of Viet Cong main force units. Village guerillas routinely passed intelligence on events in and around their village to district and provincial authorities within the Viet Cong organization. This type of reliable, systematic intelligence helped the insurgency maintain its grip on the rural villages that were their power base.[8]

Mass communications such as radio can also be used for collecting and passing intelligence to the insurgency. The Karen National Liberation Army in Burma used Radio Kawthulay in the 1980s and 1990s to provide casualty reports and replay information to the diaspora in Thailand. They also used captured VHF radios recovered during combat operations with the Burmese military forces to acquire tactical information about the military's operations. In Sierra Leone, the Revolutionary United Front (RUF) likewise acquired information about United Nations operations through captured radios and also exploited the BBC Africa service as an unwitting participant in the passage of mission orders to RUF commanders in the field, including instructions to converge on the capital of Freetown for "Operation No Living Thing" in the late 1990s.

SABOTAGE INTELLIGENCE

Sabotage is an underground function that aims at destructive attacks on critical and vulnerable infrastructure or personnel in an attempt to weaken government control and legitimacy. As with guerilla operations, saboteurs operate in small, vulnerable groups and require exacting intelligence in order to complete their tasks.

Reconnoitering transportation and communication facilities prior to sabotage attacks occupied much of the time of French resistance persons during World War II. Often working closely with Allied advisers, these people surveyed targets earmarked for sabotage on D-day. In reconnoitering a bridge, for example, resistance members looked for such factors as the guard system covering the bridge and the bridge's construction. If a number of permanent troops were evident, a step to eliminate them had to be included in the sabotage plan; when there was only an occasional patrol, the resistance would time an attack to avoid the patrol. It was important to evaluate the bridge's construction so that the size of the explosives could be calculated. By determining the schedule of enemy train movements, saboteurs were able to destroy a stretch of railroad track while it was in use, thereby compounding the wreckage and complicating repair work. Danish railroad saboteurs had an elaborate system to provide this information. Throughout Jutland, underground members were stationed near major terminals to note the departures of enemy troop trains. Whenever one was seen, the observer telephoned prearranged code phrases to the sabotage cell in the town next on the railroad line. Members of this cell then proceeded to predetermined spots on the tracks to lay their mines. With this advance notice, the mines could be placed at the last moment, preventing detection by patrolling guards. The train delayed by sabotage might eventually reach the next stop, but there observers would be waiting to repeat the process. Using these observers along a train's route, the resistance was sometimes able to slow a train's progress by days or even weeks. In Kosovo, similar tactics and information cueing were used to target convoys along the main Pristina-Belgrade highway and effectively control this main route by placing mines and using snipers to take out key vehicles and limit the mobility of Serbian forces.[9]

Production facilities can also be surveyed by undergrounds in preparation for sabotage attacks. When possible, outside intelligence experts can aid underground personnel in planning factory sabotage, for these experts are best qualified to make the necessary technical

49

judgments: it is a problem in itself to determine just which components within a plant should be incapacitated. Prior to the blowing up of a Norwegian heavy-water plant being operated by the Germans during World War II, the preliminary reconnaissance was done by a Special Operations Executive agent parachuted into Norway. Details about the factory's equipment were obtained from a Norwegian scientist in London. Other data, perhaps about the guard system and access to the equipment, apparently were supplied by underground workers in the plant.

Attacks on national infrastructure have also become common acts of sabotage by insurgencies and require expert knowledge on which key pieces of infrastructure are most important, most expensive, and most difficult to repair. The Fuerzas Armadas Revolucionarias de Colombia (Revolutionary Armed Forces of Colombia, or FARC) and Sendero Luminoso made frequent use of attacks on electrical grids, telecommunications, and even roads and bridges in order to undermine the legitimacy of the governments they were opposing. The targets they selected were chosen in order to have the greatest impact upon the civilian populations in major urban areas and to cause significant economic impact.

SCIENTIFIC AND TECHNOLOGICAL INTELLIGENCE

Secret scientific and military data may be obtained by recruitment of employees at scientific and military installations, or by simple observation elsewhere. An example of the latter was the valuable data about the V-2 rockets obtained by the Danish resistance. During the summer of 1943, fishermen near the island of Bornholm began to report rashes of unidentified objects falling into the sea, and they were recorded by the resistance leader on the island. In August the island's police commissioner notified the underground leader of the crash of a flying craft in a nearby field. The two men rushed to the scene before the Germans and found the wreckage of what was clearly a new kind of aircraft. The only identification mark was a number: "V1-83." The men took photographs immediately before the German investigators arrived. From the skid marks, the underground leader was able to determine that the device had come from the southwest. From the photographs, the underground chief drew a complete sketch of the weapon. This sketch, the photographs, and the notations as to the direction from which the missile had come were sent by courier to England, providing the British with perhaps their first technical data on the new German rocket.

POLITICAL INTELLIGENCE

Insurgent and resistance movements are keenly interested in political developments throughout the populace of their respective countries because their success or failure depends, ultimately, on political success.

Underground agents collect political intelligence. They note the statements and activities of persons to determine who favors the regime, so that these persons may be closely watched or eliminated if their actions seriously threaten the underground. In Belgium during World War II, the resistance kept files on collaborators and campaigned by using threatening phone calls and letters to dissuade these individuals from working with the enemy. If this failed, the collaborators were often assassinated. The list of collaborators was never made public in order to keep concealed the extent of cooperation with the enemy.

The Orange Revolution provides an excellent example of modern political intelligence and its ability to direct and fuel the actions of an insurgency. On November 21, 2004, the presidential runoff election was held in Ukraine with nonpartisan exit polling showing the challenger Viktor Yushchenko with a 52 percent lead to the incumbent Viktor Yanukovich's 43 percent vote count. When the official results were released, however, Yanukovich was declared the winner with 49.5 percent to 46.6 percent for Yushchenko. More specifically, the ability to rapidly acquire and process the nonpartisan election data—and distribute the results on the Internet and via opposition radio and television—allowed the members of this revolutionary movement to pinpoint in which specific districts the voting had been rigged, thus providing the clear evidence required to mobilize the masses and garner international support.[10]

In wartime, the underground can also note the morale of the enemy soldiers. The Polish Home Army systematically collected data on German troops by reading their mail because there were too few Germans to handle all of the postal work. These workers would open letters and photograph the contents before sending them on. From these letters a fairly good estimate was made of the enemy's morale.

CONCLUSION

Insurgent and revolutionary movements subsist on good intelligence. The gathering and communication of intelligence is one of the fundamental activities of the underground, and without it such movements inevitably fail. While the collection of intelligence thus remains a critically important requirement for the organization, it is also one of the typically inherent strengths of undergrounds. An effective insurgency or revolutionary movement develops close ties with the population—rural, urban, or both. The closer and stronger those connections are, the better the flow of relevant, timely, and accurate intelligence.

Many factors frame and influence the intelligence process. Language, cultural affinity, and methods of communications can either facilitate or inhibit intelligence efforts. The intelligence process is so closely linked with an insurgency's success or failure that it becomes a key focus of successful counterinsurgency. Government countermeasures, particularly when informed by enlightened counterinsurgency strategy and reinforced with effective governance that aims at solving core issues, can disrupt and neutralize an insurgency's intelligence efforts. These countermeasures include a spectrum of activities that ranges from simple operational security practices to the detection, identification, and elimination of enemy agents. When an insurgent group is denied effective intelligence, it cannot operate effectively, and its vulnerability to countermeasures increases. Absent relevant information on government activities, forces, dispositions, and weaknesses, such groups must either hazard ill-advised attacks or lapse into inactivity, threatening the movement's growth and relevance. Hence, effective counterinsurgency must include strong measures to disrupt enemy intelligence.

ENDNOTES

1   Bryan Gervais and Jerome Conley, "Viet Cong: National Liberation Front for South Vietnam," in *Assessing Revolutionary and Insurgent Strategies*, ed. Chuck Crossett (Laurel, MD: The Johns Hopkins University Applied Physics Laboratory, 2009), 13.

2   Ron Buikema and Matt Burger, "New People's Army (NPA)," in *Casebook on Insurgency and Revolutionary Warfare, Volume II: 1962–2009*, ed. Chuck Crossett (Laurel, MD: The Johns Hopkins University Applied Physics Laboratory, 2010), 4–27.

3   Summer Newton, "Provisional Irish Republican Army (PIRA)," in *Assessing Revolutionary and Insurgent Strategies*, ed. Chuck Crossett (Laurel, MD: The Johns Hopkins University Applied Physics Laboratory, 2009), 37.

4   Michael J. Deane and Maegen Nix, "Liberation Tigers of Tamil Eelam (LTTE)," in *Assessing Revolutionary and Insurgent Strategies*, ed. Chuck Crossett (Laurel, MD: The Johns Hopkins University Applied Physics Laboratory, 2009), 56.

5   Deane and Nix, "Liberation Tigers," 55.

6   Chuck Crossett and Summer Newton, "The Provisional Irish Republican Army: 1969–2001," in *Casebook on Insurgency and Revolutionary Warfare, Volume II: 1962–2009*, ed. Chuck Crossett (Laurel, MD: The Johns Hopkins University Applied Physics Laboratory, 2009).

7   Andrew R. Molnar, *Undergrounds in Insurgent, Revolutionary, and Resistance Warfare* (Washington, DC: Special Operations Research Office, The American University, 1963), 108.

8   Gervais and Conley, "Viet Cong," 31–32.

9   Maegen Nix and Dru Daubon, "Kosovo Liberation Army, 1996–1999," in *Casebook on Insurgency and Revolutionary Warfare, Volume II: 1962–2009*, ed. Chuck Crossett (Laurel, MD: The Johns Hopkins University Applied Physics Laboratory, 2009).

[10] Jerome Conley, "Orange Revolution (Ukraine): 2005–2005," in *Casebook on Insurgency and Revolutionary Warfare, Volume II: 1962–2009*, ed. Chuck Crossett (Laurel, MD: The Johns Hopkins University Applied Physics Laboratory, 2009).

CHAPTER 4.

FINANCING

CHAPTER CONTENTS

Robert Leonhard and Jerome M. Conley

INTRODUCTION

Nothing can happen in insurgent movements without money. Financing revolutionary movements is a key underground function, and the very methods of obtaining funds have a direct impact on the nature, ideology, and strategy of the movement. It is conceivable that a spontaneous resistance movement might emerge from public dissatisfaction, a protest march, or a riot, but sustaining and growing a movement into something that will effect change requires time, patience, and above all, money.

Depending on their activities, undergrounds may need money to meet the following expenses: the salaries of full-time workers in the organization; advances of money to persons who need money to pay contacts or buy food while traversing an underground escape route; the purchase of materials, such as Internet access, for propaganda publications; the purchase of explosives and other materials for sabotage; and the purchase of communications equipment, etc. An underground may also extend aid to families who shelter refugees to enable them to buy extra food. This happened in Belgium during World War II after the Nazis eliminated many resistance collaborators from the bureaucracy. Previously, these sympathizers supplied fugitives with documents enabling them to switch identities and hold jobs. When this source of papers no longer existed, it was necessary for many evaders to go into hiding. Money to care for them was supplied by the treasury of the *Armee de Belgique.*

Financial aid may be extended to the families of underground workers who have been captured or forced to flee. Typical of this was the support given by the Luxembourg resistance to the dependents of 4,200 persons who were deported and nearly 4,000 who were sent to prisons and concentration camps during the Nazi occupation. *L'Oeuvre Nationale de Secours Grande-Duchesse Charlotte* not only provided immediate care for orphans but also gave each a 30,000-franc trust fund. At the same time in Belgium, *Fonds de Soutien* (Funds for Support) was begun by the *Mouvement National Belge* for the families of workers in hiding. Similarly, Ḥarakat al-Muqāwamah al-'Islāmiyyah (Islamic Resistance Movement, or HAMAS) and other Palestinian resistance movements would offer families of suicide bombers or other "martyrs" financial compensation as a means of encouraging other impoverished families to produce fighters for the cause.

Money is also needed for bribery. Insurgencies thrive under corrupt governments, and undergrounds often disperse money to key officials to obtain their protection or silence. Bribery also plays a part in subversion and the gathering of intelligence. In most areas of the world, money equates to power and influence.

An underground may also channel funds to military units to pay salaries and buy supplies. In the Philippines, it was a prime responsibility of the Communist Politburo in Manila to obtain money for the Hukbalahap movement; and in Malaya, the *Min Yuen* was the major supplier of money to the rebels, obtaining many funds by extorting money from large landowners and transportation companies and by appropriating cash from Communist-dominated unions. Hizbollah in Lebanon required money to maintain a sizable guerilla force, along with missiles aimed at Israel. In a similar manner, undergrounds that support a guerilla component require reliable funding sources.

Some insurgencies require funds in order to support their social outreach work and shadow government activities. Just as legitimate state governments struggle with the rising cost of medical care, unemployment insurance, food aid, housing subsidies, and pensions, some insurgent movements also struggle to provide similar services in an attempt to undermine the government, care for their constituents, and provide a cover for illegal and violent activities. These activities are expensive and require sustained and reliable income.

The cost of high-profile individual acts of violence tends to be miniscule compared with the overhead costs of simply sustaining and administering the underground, auxiliary, armed component, and public component. Spectacular acts of terror can inspire increased devotion and donations from supporters for a relatively low cost. But the day-to-day administration required to sustain and grow a movement is considerable.[1] Undergrounds obtain financing through a combination of external and internal sources.

EXTERNAL SOURCES OF MONEY

Foreign Governments

Often an underground is aided by an outside sponsor, usually a government, but more recent examples include non-state actors. Much of the money used by the anti-Nazi Belgian resistance of World War II, for example, came from franc reserves in London released by the British government. At one time, 10 million francs per month were forthcoming. Similarly, many of the funds used by the French resistance were remitted from the Bank of England or sent from the Bank of Algiers (after the Allied landing in North Africa). The Viet Cong's resistance against the government of South Vietnam and its American allies was funded by both China and the Soviet Union, as well as by the authorities in Hanoi. Some external sponsors, such as Colonel Muammar Gaddafi in Libya, support a number of different insurgencies. Gaddafi provided funding to groups ranging from Charles Taylor and the Revolutionary United Front (RUF) in Liberia and Sierra Leone, respectively, to the Provisional Irish Republican Army (PIRA) in Northern Ireland. Other sponsors, such as Iran, however, tend to focus their sponsorship funding on insurgencies that advance the Shiite cause and those insurgencies that undermine Israeli (and Western) influence, such as the Palestinian cause.

Foreign governments extend support to undergrounds for several reasons. The most important is that the activities of an underground often contribute to the defeat of a common enemy. Such aid also enables the sponsor to demand some reciprocity on the part of the underground. However, an outside government may give financial assistance to an underground even if there is no common enemy. According to one report, such a case occurred in 1940 when the Japanese government—not yet allied formally with Germany and Italy—provided the Polish underground in Rome with financial aid as well as technical equipment and Japanese passports in exchange for intelligence data on the German and Soviet occupying forces.[2]

Non-State Actors

In addition to governments, friendship societies or quasi-official aid groups may channel funds to an underground. Perhaps the best known of the latter was the Jewish Agency, which, during the Palestine revolution, had offices or representatives in every part of the Western world. In the run-up to the Israeli War for Independence, Palestinian Jews obtained critically needed financing from fellow Jews throughout the world, especially in Europe and the United States. Open appeals for money were made in newspapers and lectures and at charity balls and other social events.[3]

The vast majority of the Liberation Tigers of Tamil Eelam (LTTE) financing came from the large Tamil expatriate community, especially those contingents in Western countries (Canada, the United Kingdom, Australia, the United States, and Scandinavia) but also those living in the Indian province of Tamil Nadu. Indeed, many analysts identify the overseas Tamil communities as the single most important actor enabling the insurgency. Expatriate support included voluntary contributions from individuals and Tamil-owned businesses, as well as funds extorted from expatriates. LTTE collection methods evolved over time, from poorly coordinated, often violent acts of coercion to scheduled collections based on computerized databases that allowed overseas collectors to avoid paying visits to individuals who supported rival Tamil groups or who were already regular contributors. Collections were made monthly or annually; additional collections were made according to special dates commemorating specific battles or individual "martyrs." Information was also collected on extended families residing elsewhere in order to lend credibility to threats in the event that a donation was not forthcoming. A 2009 Canadian intelligence report revealed that the community of expatriates and sympathizers within Canada was one of the top contributors to the LTTE, with donations of approximately $12 million per year. After the LTTE lost control of the Jaffna Peninsula in the mid-90s, this source of financing became increasingly important, by some accounts providing up to 90 percent of the group's operating funds. However, the classification of the LTTE as a terrorist group by most Western countries, the result of an intense lobbying effort by anti-LTTE forces, put a serious strain on LTTE's ability to raise and transfer expatriate funds and was probably a major contributor to the Tigers' defeat in 2009.[4]

The main source of overseas funding for the New People's Army (NPA) in the Philippines during the Ferdinand Marcos administration was from humanitarian organizations, including a number of European churches, and radical groups in Europe. The Communists, working through their public face, the National Democratic Front (NDF), touted the NDF as the only viable opposition to the human rights abuses of the Marcos regime. By 1987, the NDF had attracted the support of numerous international human rights organizations and established support networks in more than twenty-five countries. The NPA strategy was to divert resources from the nongovernmental organizations through aboveground institutions that were run by NPA supporters under the auspices of rural aid and development. These organizations remained a major source of support even after the fall of Marcos.[5]

PIRA likewise relied on support from abroad. They turned to the United States for money and weapons as soon as they were organized enough to send agents abroad, and the Irish communities of Boston and New York proved very supportive. In 1969, the United States had five times as many Irish as Ireland. The Irish Northern Aid Committee (NORAID) was set up in New York City in 1970 to provide a steady stream of money to the IRA, mostly for the purchase of weapons.[6]

Charities and nonprofit organizations are attractive sources of funding for insurgent undergrounds because they tend to be less regulated and scrutinized than publicly owned corporations, and some have a global reach and presence, administering considerable sums of money. Insurgencies can abuse charities through fraud—e.g., appropriating monies from a real charity intended for charitable purposes and using them to fund the movement instead; through a sham organization—i.e., creating an entirely bogus charity that poses as legitimate; or through co-opting a charity's money—i.e., using the money for the purpose intended by the donators but administering and distributing those funds through the insurgency.[7]

Cash in the Local Currency

Aid is often given in the form of cash in the local currency, which has the advantage of being easily exchanged for goods or services. The main problem is the physical transfer of the money. Usually this is handled by a front business organization, through diplomatic channels, through clandestine couriers, or by infiltrated agents. A growing concern and challenge for counterinsurgency organizations, however, is the expanding ability to transfer money internationally using informal funds transfer (IFT) systems, such as the "hawala" system in the Middle East, fei-ch'ien in China, hundi in India, and padala in the Philippines, which can provide cash in the local currency for the recipient. This is discussed in more detail below under *Parallel Financial Systems*.

Substitute Currency

Hard currency, such as U.S. dollars or British pounds, is sometimes given to an underground when the sponsoring government lacks adequate reserves of the local currency. Hard currency makes a good substitute because it is easily exchanged on the black market for local currency or goods. Hard currency is also useful when the local currency is confiscated by the authorities and replaced by scrip, a frequent government countermeasure. This was done by the Castro regime soon after the Cuban revolution.

Dollars were used extensively in financing World War II undergrounds. One British agent in Yugoslavia reported that it was no trouble to use dollars (or gold pieces) because "there was invariably a market for 'good' money in the towns."[8] More recent developments include the potential use of electronic credits accrued and traded via mobile phones as a means of transferring funding from one location to the next, especially in East Africa where the use of this type of electronic currency is commonplace, and the acquisition of cash value for online video-gaming credits that can be acquired and traded with little visibility.[9, 10]

Counterfeit Money

One other way to finance an underground movement is through the use of counterfeit money. Although production of such money is not exclusively the province of a sponsoring government, undergrounds usually lack the necessary facilities and technical competency to counterfeit money; therefore, the main effort is generally undertaken by friendly governments.

Of course, the use of counterfeit money adds to the dangers already facing underground members. During World War II, this factor reportedly prompted the Polish state underground to reject an offer of counterfeit money from London.

Online Fraud

Insurgencies increasingly use illegal online operations to steal money or goods. Techniques include credit card and online banking fraud. In some cases, insurgents purchase stolen credit card or bank account numbers and passwords from criminal organizations and then use that information to withdraw money from compromised accounts or to buy goods directly. This avenue of obtaining funds illegally features an ongoing conflict between insurgents exploiting vulnerabilities in global computer networks and various government and international organizations attempting to fix those vulnerabilities and shut down online fraud through technical means, legislation, and enforcement.

Parallel Financial Systems

Islamic history and culture gave rise to an innovative and effective approach to financing insurgency as practiced by members of the Muslim Brotherhood and Al Qaeda. Hasan al-Banna, the founder of the Muslim Brotherhood, viewed finance as a critical weapon in undermining the infidels and reestablishing the Islamic caliphate. To do so, he believed Muslims must create an independent Islamic financial system that would parallel and later supersede the Western economy.[11] Al-Banna's successors set his theories and practices into motion, developing uniquely Islamic terminology and mechanisms to advance the Brotherhood's system of faith, as well as their unique financial apparatuses.

President Gamal Abdel Nasser negated the Brotherhood's attempt to establish an Islamic banking system during the mass arrests in 1964. But Saudi Arabia welcomed this Egyptian dissident idea, and in 1961, King Saud bin Abdul Aziz funded the Brotherhood's establishment of the Islamic University in Medina to proselytize their fundamentalist

Islamic ideology. In 1962, the Brotherhood convinced the king to launch a global financial joint venture that established numerous charitable foundations across the globe. This joint venture became the cornerstone of the Brotherhood and was used to spread Islam (and later to fund terrorist operations) worldwide. The first of these charitable organizations were the Muslim World League and Rabitta al-Alam al-Islami, which united Islamic radicals from more than twenty nations. In 1978, the kingdom backed another Brotherhood initiative, the International Islamic Relief Organization (IIRO), an entity that has been implicated in funding organizations such as Al Qaeda and HAMAS.

Most Muslim nations collect mandatory Islamic charity (zakat) of approximately 2.5 percent from Muslim institutions and companies.[12] Zakat is intended to go to those who are less fortunate. However, the Brotherhood determined that those engaged in jihad against the enemies of Islam are entitled to benefit from the charitable offering. The interpretation that modern jihad is a serious, purposefully organized work intended to rebuild Islamic society and state and to implement the Islamic way of life in the political, cultural, and economic domains is widely accepted among Muslims, and thus those involved in jihad are viewed as legitimate recipients of zakat.[13]

The rise of global militant Islam, and Al Qaeda in particular, also benefited from the use of parallel financial systems. The roots of Al Qaeda lie in a number of different organizations, including the *Maktab Al Khidamat* (MAK) or Services Office, a clearinghouse established to facilitate the recruitment, transportation, organization, training, and equipping of Arabs to support the Afghan resistance. Established by Abdullah Yusuf Azzam, a Palestinian scholar of Islamic law, and Osama bin Laden in Peshawar, Pakistan, in 1984, the MAK also attracted other militant leaders.[a] The MAK consisted of a network of international recruiting offices, bank accounts, and safe houses and was also responsible for the construction of paramilitary camps for the training of militants and the fortifications used by Arab fighters.[14] Between 1982 and 1992, estimates report approximately 35,000 foreign fighters contributed to the Afghan effort, although there were probably never more than 2,000 in Afghanistan at any one time. The MAK was

---

[a] Azzam, who earned a Ph.D. from Cairo's Al-Azhar University, was a very charismatic character and played a central role in crafting the narrative of resistance that drew thousands of Arabs to the Afghan cause. His previous combat experience (he fought the Israelis in 1967), combined with his religious credentials (his education and his connections with the family of Sayyid Qutb, an important ideological leader of the early Muslim Brotherhood), made him a particularly appealing figure to bin Laden. Omar Abd Al-Rahman, an Egyptian militant also educated at Cairo's Al-Azhar University and a key ideological figure for both Al-Jama'at Al-Islamiyya and the Egyptian Islamic Jihad, also utilized MAK's resources to contribute to the Afghan resistance.

responsible for training approximately 12,000–15,000 of those fighters, with approximately 4,000 remaining connected through either chain of command or ideological affinity after the conflict.[15] To complicate the tracking of finances, leaders employed parallel financial systems drawn from historical Islamic roots.

Both the Egyptian Islamic Group (EIG) and the Egyptian Islamic Jihad (EIJ) likely utilized these mechanisms to their advantage. EIG often sought donations during Friday prayers at mosques in order to facilitate their social outreach programs. It is unknown how much of this funding went toward illicit activity; however, it is presumed that the preponderance of the collected funds was reinvested into public and legal organizational activities. The extent of the aid EIJ received from outside of Egypt is not known, although the Egyptian government has claimed that both Iran and Saudi Arabia have provided financial and material support to EIJ.[16]

Given the prevalence of obtaining funds through such means, it may be assumed that EIJ obtained some funding through various Islamic nongovernmental organizations, cover businesses, zakat funding operations, and possibly, although not likely, criminal acts.[17] The most evident external support to EIJ was its symbiosis with Al Qaeda and its increasing dependence on that organization. Few exact figures exist; however, from 1996 to 1997, EIJ received more than $5,000 per month[b] from Al Qaeda.

In the hawala system, money is transferred from a worker in one country to another worker in another country through the use of intermediary "hawaladars" in each country. Although a fee (cash, goods, or services) is charged for this transaction, the fee is usually much less than that charged in the formal banking sector, and this system also allows for the transfer of funds to/from countries and regions with limited financial infrastructure.[18]

---

[b] In late 1996, Dr. Ayman Al-Zawahiri traveled clandestinely to a number of former Soviet Caucasian republics (including Chechnya) and in December was arrested by a Russian patrol in Dagestan. He was released in May, having stuck to his cover story without being identified by the Russians; however, he was chastised by Al Qaeda members for his carelessness and saw the subsidy (paid to EIJ by Al Qaeda) lowered to $5,000 during his imprisonment.

INTERNAL SOURCES OF MONEY

Noncoercive Means

Gifts

Voluntary gifts from wealthy individuals and, occasionally, from commercial enterprises have constituted a good source of income for many undergrounds and are easier to hide from security forces. A few wealthy Chinese businessmen in Manila made large gifts to the Hukbalahap; the Malayan *Min Yuen* received substantial aid from several Chinese millionaires in Singapore. Many industrialists and bankers provided funds for the anti-Fascist underground in Italy. However, donor firms in France during the resistance encountered difficulties in hiding their donations from the Germans, and this hampered the exploitation of this potential source of revenue. In addition, financial gifts to the underground can also come from friends and relatives of underground workers and, given the manpower and opportunity, an underground may canvass door to door for contributions. Finally, dues levied on underground members can also provide needed funds.

Loans

The underground may also borrow funds. The Yugoslav Partisans, for example, floated a 20-million-lira loan, which was marketed among the Slovene populace as "Liberty Loans"; and Belgian banker Raymond Scheyven's Service Socrates organization managed to borrow in the name of the government-in-exile more than 200 million francs for the anti-Nazi Belgian underground from the end of 1943 to liberation.

A problem that sometimes confronts an underground worker in soliciting funds from strangers is that of convincing them of the agent's good faith. The underground worker may provide strangers with an official-looking document authorizing him to collect funds and sign notes. The Service Socrates used a more complicated system, however. This organization invited prospective lenders to suggest a phrase to be mentioned on the BBC on a given night. The underground passed the requests on to the London authorities, the phrase was broadcast at the designated time, and the individuals knew that they were dealing with bona fide agents of the underground. To safeguard the Service Socrates and the Belgian government in London against future false claims, lenders were given certificates stating the amount of the loans and bearing a number. Raymond Scheyven, using his pseudonym, "Socrates," signed these certificates, and a copy of this signature was on file in London for comparison at the time of repayment after the war.

If the underground can borrow in the name of some constituted authority such as a government-in-exile, it is more likely to receive a favorable response than if funds are sought in the name of an aspiring underground whose trustworthiness as a debtor organization may be in doubt. As one writer expressed it, governments-in-exile provide necessary "symbols of legalism."[19]

In addition, if an underground has access to some form of collateral, such as oil distribution networks or diamond fields, they may be able to secure funding, weapons, and other needed assets in exchange for granting access to this collateral or resource. The RUF in Sierra Leone obtained funds and weapons from Liberia and Libya in exchange for diamonds and access to mines.[20]

Embezzled Funds

An underground may obtain funds embezzled from government agencies, trade unions, businesses, and nongovernmental organizations. An example is the secret appropriations that the Danish resistance received from the Royal Treasury to support the publication *Information*. Also, misappropriated grand duchy revenues constituted perhaps half of the money raised for the anti-Nazi resistance in Luxembourg. Trade union funds were embezzled on a fairly large scale by Communist leaders of Malayan trade unions in the years 1945–1947 and provided a major source of income for the Malayan Communist Party (MCP) until the British replaced the Communists with unionists loyal to the government. In Somalia, drought relief funding and supplies were interdicted by Al Shabaab to support their network, and in Rwanda, an estimated $112 million of foreign aid was used to purchase weapons—mostly machetes—from France, South Africa, and Egypt.[21]

Sales

The sale of various items through door-to-door canvassing or through "front" stores may provide money. Yugoslav Communists once sold fraudulent lottery tickets. The Luxembourg resistance sold lottery tickets as well as photographs of the Grand Duchess. In post-World War II Malaya, the MCP treasury was supplemented by funds obtained from party-owned bookstores, coffee shops, and even small general stores. Similarly, the Yugoslav Communists raised money through sales made by party-owned clothing stores.

Coercive Means

Robberies

To bring in money, undergrounds frequently resort to holdups. The Hukbalahap in the Philippines, for instance, was able to collect funds by staging train robberies. Likewise, the Organisation de l'Armée Secrète (Organization of the Secret Army/Secret Armed Organization, or OAS) in Algeria conducted a series of bank robberies. In Malaya, the Communists formed a "Blood and Steel Corps" to engage in payroll robberies and raids on business establishments. Business firms, rather than individuals, are usually the targets of such robberies. The LTTE had numerous methods by which it mobilized resources to sustain operations. These included various criminal activities, such as bank robberies, extortion, and the smuggling of drugs and other contraband, but they also included more traditional fundraising activities that also incorporated varying levels of coercion.[22]

Undergrounds generally avoid outright confiscations from the general populace for several reasons. First, widespread robberies would tend to brand an underground as an outlaw band and destroy its public image as a potential legitimate authority. Second, simple confiscations of money would not make the victims compliant servants of the underground, as other forms of coercion can do. Finally, robberies preclude the possibility of exacting continued support under the threat of exposing the affected persons' assistance to the underground.

Kidnapping and Hijacking

The practice of kidnapping in order to collect a ransom has been conducted by insurgent groups across the globe, ranging from the Fuerzas Armadas Revolucionarias de Colombia (Revolutionary Armed Forces of Colombia, or FARC) in Colombia, to the Taliban in Afghanistan, to Al Qaeda in the Islamic Maghreb (AQIM) in Northern Africa, to the Movement for the Emancipation of the Niger Delta (MEND) in Nigeria. These organizations utilize elaborate networks of middlemen and negotiators to exchange their captives for funding. During the period of 2005–2010, AQIM alone raised an estimated $65 million from kidnappings, which accounted for 90 percent of its revenue and an average rate of $6.5 million for a Western hostage.[23] The dramatic escalation in hijackings off the Somali coast and into the Red Sea and Indian Ocean has reportedly also lined the pockets of Al Shabaab insurgents who charge pirates and their villages a protection fee or tax after the ransom is paid.

Forced "Contribution"

Although undergrounds usually do not rob the public so as to avoid alienating their constituency, they sometimes coerce individuals into making donations under the tacit threat of reprisals. Aggressive application of this technique is usually reserved for wealthier persons. Typical was the practice of the OAS in Algeria, which fixed the amounts of contributions to be exacted from persons in professional occupations but allowed people of modest means to give what they wanted. A person received a typewritten note in the mail informing him that a *"percepteur"* of the OAS would call in the near future to collect his contribution. The *percepteur* was well dressed and curt but polite. If his credentials were questioned—some crooks tried to extort money in the name of the OAS—he could show a photostat message signed by the commander in chief of the OAS, General Raoul Salan. If the person refused to pay, he would not be threatened, but a week later his car or home would probably be bombed by a charge of plastic explosives. The OAS would then increase its assessment to cover the cost of the reprisal. After a few such object lessons, most of the people approached were willing to make a "contribution," and many agreed to make regular payments.

From its inception, the FARC in Colombia survived in part by obtaining funds through extortion, kidnapping/hostage-taking, and stealing supplies. At first, the FARC filled its coffers through extortion, employing tactics that were Mafia-like. The FARC would ask businesses whether they needed a "vaccination." For example, one Colombian Coca-Cola distributor decided that it did not need to be "inoculated." In response, the FARC burned forty-eight delivery trucks, kidnapped eleven workers, eventually killing seven of them, and robbed the company 400 times.[24]

The Yugoslav Partisans were able to utilize this coercive technique to their political as well as financial advantage. By exacting large amounts from landowners, the Partisans were able to weaken their political opponents, and going one step further, they eliminated some of these political competitors by denouncing them to the Germans as helpers of the underground. A check of landowners' financial records sometimes revealed unaccountable deficits, which led to the landowners' arrest and the confiscation of their properties.

An underground may suffer a setback, however, if a popular person refuses to contribute and the underground does not dare to make the usual reprisal out of fear of public indignation. A case in point was the widely publicized refusal of the French actress, Brigitte Bardot, to aid the OAS. Such a response serves to weaken the image of infallibility and complete control that the underground tries to cultivate.

Taxes

Taxes may be levied against the general public in areas where enemy forays are not frequent or serious enough to prevent underground municipal administrators from collecting taxes with the backing of nearby military units. The tax may be levied on a per capita basis, as was done in Philippine areas under Hukbalahap control, or it may be levied on a more selective basis, affecting only persons with regular incomes above a certain level, as was apparently the practice in the Slovene area of Yugoslav Partisan control.

The Tamil Tigers (LTTE) raised funds domestically by levying taxes on the population, especially in the early stages of the conflict. Those who could not pay were often incarcerated, while those families who had sons or daughters serving in the cadres were exempted. Proof of payment of this tax served as a pass for traveling through LTTE-controlled territory and for serving in administrative positions. Extracting payment was relatively inexpensive; once the Tigers established their reputation for ruthlessness, few families had to be reprimanded.

For the Taliban, the decision to formalize and tax the heroin drug economy led to significant revenues for the insurgency. By 2000, Afghanistan produced three times more opium than the entire rest of the world, with 96 percent of this Afghan production taking place in Taliban-controlled territory. The Taliban collected a 20 percent tax from opium dealers as well as the drug transporters, leading to an annual tax revenue base of approximately $20 million for the Taliban by 2000.[25]

Narcotics and Black Market Trade

Modern insurgencies have increasing connections to illegal drug trade throughout the world. The burgeoning industry of supplying marijuana, cocaine, methamphetamine, heroin, and other drugs offers opportunities for financing that most underground leaders find too lucrative to ignore.

Although extortion and kidnapping sustained the FARC for many years, growing the "little guerrilla army" required a corresponding growth in funding. To do this, during the 7th Guerilla Conference, the FARC developed a plan to leverage four commodities on the black market: livestock, commercial agriculture, oil, and gold. This would not prove to be enough, however. Eventually, without departing from the extortion and kidnapping, FARC reluctantly became involved in the narcotics trade. Initially, both Manuel Marulanda and Jacobo Arenas were opposed to *las drogas* for ideological reasons. In the long run, however, pragmatism won the day. The flow of illicit money and goods through areas that the FARC controlled was just too rich a source to allow it to pass by. In 1998 alone, for example, armed groups involved in the marketing of illegal narcotics in Colombia had proceeds in excess of $550 million. Still, the FARC became involved piecemeal. Their first step was to tax narco-traffickers while protecting the peasant farmers who grew the coca, a practice that was not unlike their "inoculations."[26]

The estimates of FARC financing obtained through narcotics run from at least $30 million annually to as high as $1.5 billion. In fact, the FARC became so sophisticated that they developed standard costs for the drug trade that in October 1999 equated to $15.70/kilo for cocaine paste, $5,263 to protect a laboratory, and $52.60 to protect a hectare of coca, etc. At one point, the FARC was responsible for exporting 50 percent of the cocaine consumed worldwide. Whatever the figure, the revenues from the narcotic trade still represented only half of FARC funding. The Colombian government suggests that the rest came from the FARC's classic funding lines of kidnapping, robbery, and extortion.[27]

When undergrounds become involved in the drug trade or other black market activities, it tends to impact the organization's core ideology and strategy. The FARC experience, again, is illustrative. With the new drug trade came money and corruption. Some of the FARC in coca-rich areas began to live as drug lords, replete with gold jewelry, fancy cars, and other luxuries. This created dissent in the ranks as FARC members who stayed true to the guerrilla life realized others were living as gangsters. To solve this problem, the FARC leadership created the National Financial Commission. Responsible for allocation of all FARC funds, including major purchases, the commission reported directly to the central leadership. A system was developed wherein all FARC units were given a funding line and direction in how to employ it. When these measures did not completely solve the corruption issue, the senior leaders assigned *ayudantías* or "advisors" to monitor what was happening at every level and provide advice to local leaders from time to time. If they suspected any foul play, the *ayudantías* would call for an investigation. Theft or even misappropriation of FARC funding was punishable by death.[28]

The Maoist insurgency in Peru, Sendero Luminoso, likewise became involved in the illegal drug trade. By the mid-1980s, having expanded in the primary coca-producing region of Peru, the Upper Huallaga Valley, the Shining Path was able to tap the profits of the drug traffickers. In order to exploit the ever-burgeoning cocaine trade, Sendero sent units into the Upper Huallaga Valley to identify and kill government enforcement agents and their supporters. Once they had taken effective control of the region, Sendero served as middlemen between the coca growers and the drug traffickers, reportedly receiving 10 percent of the sale of every kilo of coca and earning between $20 and $50 million annually, which was used to purchase weapons and pay militants. Although Abimael Guzman originally disavowed any connection between Sendero Luminoso and the ongoing drug trafficking along the Andean ridge, pragmatic considerations appear to have triumphed. The perceived disconnect between Sendero's purported ideological purity and its involvement in the black market drug trade did impact its legitimacy, however.

CONCLUSION

Finances are the lifeblood of an insurgent or revolutionary movement. Normally such organizations obtain financing through a combination of internal and external sources. The nature of the movement's collections tends to color the organization, in some cases despite its founding ideology. When an insurgency dips into the lucrative illegal drug trade, they attract attention as part of the drug trafficking problem. When they rob or extort citizens within a country, they gain a reputation as a criminal organization. When they subsist off of substantial funding from a foreign power, they tend to be viewed as a puppet organization taking its cues from that distant power.

Likewise, how the movement handles acquired funds tends to characterize the organization in the eyes of the indigenous population. Insurgencies that distribute funds to impoverished citizens gain favor as the champions of the underprivileged. Conversely, leaders who succumb to the temptation of corruption tend to bring discredit on their organizations.

Thus, the function of financing remains a key and defining activity for undergrounds. Wise insurgent leaders think of financial matters as part and parcel of the overall strategy and ideology of the movement, rather than as a peripheral matter. Likewise, government forces conducting counterinsurgency must examine sources and management of insurgent finances as part of their overall strategy to defeat the movement.

ENDNOTES

[1] Financial Action Task Force/Groupe d'action financiere, "Terrorist Financing" (Paris: OECD, 2008), 7–10, http://www.fatf-gafi.org/dataoecd/28/43/40285899.pdf.

[2] Andrew R. Molnar, *Undergrounds in Insurgent, Revolutionary, and Resistance Warfare* (Washington, DC: Special Operations Research Office, The American University, 1963), 62.

[3] Ibid.

4 Michael J. Deane and Maegen Nix, "Liberation Tigers of Tamil Eelam (LTTE)," in *Assessing Revolutionary and Insurgent Strategies*, ed. Chuck Crossett (Laurel, MD: The Johns Hopkins University Applied Physics Laboratory, 2009), 52–53.

5 Ron Buikema and Matt Burger, "New People's Army (NPA)," in *Casebook on Insurgency and Revolutionary Warfare, Volume II: 1962–2009*, ed. Chuck Crossett (Laurel, MD: The Johns Hopkins University Applied Physics Laboratory, 2010), 104.

6 Summer Newton, "The Provisional Irish Republican Army: 1969–1998," in *Assessing Revolutionary and Insurgent Strategies*, ed. Chuck Crossett (Laurel, MD: The Johns Hopkins University Applied Physics Laboratory, 2009), 64–66.

7 Financial Action Task Force, "Terrorist Financing," 11–12.

8 Molnar, *Undergrounds*.

9 Darlene Storm, "Intelligence Agencies Hunting for Terrorists in World of Warcraft," *ComputerWorld* (blog), April 13, 2011, http://blogs.computerworld.com/18131/intelligence_agencies_hunting_for_terrorists_in_world_of_warcraft.

10 "Al-Shabab Bans Mobile Phone Money Transfers in Somalia," *BBC Online*, October 18, 2010, http://www.bbc.co.uk/news/world-africa-11566247.

11 Rachel Ehrenfeld, "The Muslim Brotherhood New International Economic Order," *The Terror Finance Blog* (blog), October 13, 2007, http://www.terrorfinance.org/the_terror_finance_blog/2007/10/the-muslim-brot-1.html.

12 Ibid.

13 Ibid.

14 Lawrence Wright, *The Looming Tower: Al-Qaeda and the Road to 9/11* (New York: Vintage Books, 2006). Azzam originally established the MAK—and later persuaded bin Laden to join. Bin Laden used his family's relationship with the Saudi Royal Family to support the effort overtly—through a strategic communications plan—and covertly, eventually matching U.S. financial contributions to the resistance.

15 Peter L. Bergen, *Holy War, Inc.: Inside the Secret World of Osama Bin Laden* (New York: The Free Press, 2001).

16 Michael Scheuer, *Through Our Enemies' Eyes: Osama Bin Laden, Radical Islam, and the Future of America* (Washington, DC: Potomac Books, Inc., 2008).

17 Marc Sageman, *Understanding Terrorist Networks* (Philadelphia: University of Pennsylvania Press, 2004).

18 Mohammed El-Qorchi, "The Hawala System," *Finance and Development* 39, no. 4 (December 2002), http://www.gdrc.org/icm/hawala.html.

19 Molnar, *Undergrounds*, 64.

20 Jerome Conley, "The Revolutionary United Front (RUF)," in *Assessing Revolutionary and Insurgent Strategies*, ed. Chuck Crossett (Laurel, MD: The Johns Hopkins University Applied Physics Laboratory, 2009), 52–53.

21 Bryan Gervais, "Hutu-Tutsi Genocides," in *Casebook on Insurgency and Revolutionary Warfare, Volume II: 1962–2009*, ed. Chuck Crossett (Laurel, MD: The Johns Hopkins University Applied Physics Laboratory, 2009).

22 Deane and Nix, "Liberation Tigers," 52–53.

23 Jean-Charles Brisard, "AQIM Kidnap-for-Ransom Practice," *The Terror Finance Blog* (blog), September 27, 2010, http://www.terrorfinance.org/the_terror_finance_blog/2010/09/aqim-kidnap-for-ransom-practice-a-worrisome-challenge-to-the-war-against-terrorism-financing.html.

24 Stephen Phillips, "Fuerzas Armadas Revolucionarias De Colombia—FARC," in *Assessing Revolutionary and Insurgent Strategies*, ed. Chuck Crossett (Laurel, MD: The Johns Hopkins University Applied Physics Laboratory, 2009), 23.

25 Sanaz Miraz, "Taliban 1994–2009," in *Casebook on Insurgency and Revolutionary Warfare, Volume II: 1962–2009*, ed. Chuck Crossett (Laurel, MD: The Johns Hopkins University Applied Physics Laboratory, 2009), 471–472.

[26] Phillips, "FARC," 24.

[27] Ibid., 24–25.

[28] Ibid., 25.

CHAPTER 5.

LOGISTICS

CHAPTER CONTENTS

Robert Leonhard and Jerome M. Conley

INTRODUCTION

In military usage, logistics includes the critical functions of procuring, storing, and distributing supplies. It also includes maintenance, medical services, and transportation. Military supplies include food, water, general supplies, fuel and oil, building materials, ammunition, major end items like weapons and vehicles, medical supplies, and repair parts. As an insurgent or revolutionary movement grows—and particularly as it develops a guerilla component—the underground must acquire the skills to manage logistical support.

Logistical operations are required to meet the materiel demands of both the underground and the guerrilla forces. Because underground workers are generally engaged in civilian occupations, they are usually able to provide their own basic supplies of food, clothing, and medicines. What they need are operational supplies—printing equipment, paper, ink, radios, and sabotage implements. Guerrilla logistical needs, including food, clothing, medicines, arms, and munitions, are both basic and operational, and these forces have usually relied in part on underground logistical operations to provide such supplies.

In practice, logistical functions are shared by both the underground and the auxiliary. Specific procedures vary according to the context of each insurgency or resistance, but in general the auxiliary handles routine logistics—especially food, water, and fuel—while the underground often takes on the more difficult task of procuring and distributing large-caliber ammunition or other special supplies. The underground personnel generally plan and supervise logistical functions, relying on the auxiliary personnel when possible.

PROCUREMENT

Purchases

Black Market

Undergrounds sometimes purchase supplies on the black market—from persons who own or have access to certain goods and who are willing to sell or trade those goods in spite of legal restrictions. For instance, some workers in an Italian anti-Fascist underground had the specific assignment of bartering with a black market sponsored by some young Fascists. This market flourished during a period when the demand for staple goods was very high. Reportedly, 220 pounds of salt could be exchanged for an excellent machine gun.

The experiences of the Provisional Irish Republican Army (PIRA) are instructive regarding how insurgencies obtain weapons, ammunition, and other supplies. The quartermaster general, a member of the PIRA leadership council, had the duty of procuring, transporting, and storing weapons. The quartermaster often had regional quartermasters. Weapons were strictly controlled. Bunkers provided storage, and weapons were supposed to be dispensed solely for operations. In some areas, small numbers of weapons were held by the local service units for guard duty or small operations, but on the whole the disciplined control of arms was maintained.

The lack of arms for defense was one of the major dissatisfactions with the original IRA leadership in 1969 that led to the breakaway Provisional movement. Therefore, arms were in short supply during the early years, and the leaders concentrated much of their initial efforts on pressuring the Republican-leaning movement for cash and arms. Firearms were restricted in Northern Ireland, so the Provisionals knew they had to establish a major flow into the country from outside sources. Early in the campaign, mines and explosives were also in short supply for the PIRA. Gelignite was stolen from local mines to make the first explosives, and then fertilizers were used until the arrival of Semtex from Libya. The success of the British army's defusing capability, however, led the group to move from simple timers to more complex detonation devices that included anti-handling features. The PIRA also introduced the car bomb, which allowed a larger explosive charge and a more discreet and safer emplacement as well as the ability to command detonations using remote-control transmitters.

The eventual flow of arms into Northern Command operational areas arrived mostly via Southern Command routes, which included safe houses, staff offices, caches, storage, and transportation. PIRA members were sent abroad to the United States to collect money and any arms they could get. By 1972, shipments of military machine guns were arriving from the United States. Soon thereafter, multiple types of arms and explosives started to flow from Libya. One network alone shipped hundreds of light, collapsible, concealable ArmaLite rifles during the 1970s. The British security forces confiscated more than 700 weapons, 2 tons of explosives, and more than 150,000 rounds of ammunition in 1971 alone, most of which came from the United States. The fortunes of the PIRA later depended on the relationship they established with Colonel Muammar Gaddafi in 1972. The head of Libya saw himself as a catalyst for revolutionary movements around the world and agreed to meet with an IRA representative. Although shipments began to flow to Dublin in 1973, they were often interdicted in high-profile seizures. The *Claudia* was boarded by Irish authorities to reveal that the IRA was willing to accept large amounts of money and weapons from the state-sponsor. This was followed by the *Eksund*, which was captured by French authorities on November 1, 1987, with over 150 tons of armaments: 1,000 AK-47s, 1 million rounds of ammunition, 430 grenades, 12 rocket-propelled grenade launchers, 12 machine guns, more than 50 SA-7 surface-to-air missiles (SAMs), 2,000 electric detonators, 4,700 fuses, 106-mm cannons, anti-tank missiles, and 2 tons of Semtex.[1]

Black market weapons have seriously impacted sub-Saharan Africa and aggravated conflicts throughout the continent. Insurgencies, along with other state and non-state actors, have multiplied armed conflicts across the African continent since the 1960s, often on the shoulders of illegal arms merchants, who in turn profit from selling small arms and other munitions to all sides in a conflict. Africa leads the world in the number of armed conflicts, and in every case these wars have been exacerbated by illegally trafficked arms—weapons that have claimed over 7 million lives on the continent. The devastation wrought by arms merchants in Africa contributes to the epidemic of displaced persons, crimes against humanity, and the perpetuation of societal violence that spawns radicalization among war-ravaged youth. Illegal arms trafficking sustains conflicts in Angola, Burundi, Chad, Cote d'Ivoire, Democratic Republic of Congo (DROC), Djibouti, Eritrea-Ethiopia, Guinea, Guinea-Bissau, Kenya, Liberia, Nigeria-Cameroon, Republic of Congo, Rwanda, Senegal, Sierra Leone, Somalia, Sudan, Tanzania-Zanzibar, Uganda, and Zimbabwe.

Insurgents operating in weak or failed states in Africa sometimes purchase arms using stolen or otherwise misappropriated noncash resources such as diamonds, gemstones, ivory, and oil. Examples include the Congolese Liberation Front (FLC), the Mai Mai militia groups in the Democratic Republic of the Congo, and the Revolutionary United Front (RUF) in Sierra Leone. International efforts aimed at curbing the use of such resources for purchasing weapons have enjoyed only modest success, and the practice continues to flourish.

Porous borders, weak international regulation, and poorly administered trans-shipment points help the illegal arms market in Africa thrive. The failure to enforce United Nations' sanctions and embargoes leaves insurgent undergrounds relatively free to purchase arms at will. Traffickers who routinely violate such measures are rarely prosecuted, and the banks and other financial institutions that facilitate illegal arms sales generally do so with impunity. The combination of these factors tends to strengthen insurgencies, lengthen conflicts, and increase the price paid by regional noncombatants.[2]

Legal Market

Semi-finished items for manufacturing may often be purchased by undergrounds from legal firms. In most cases, this is done through a front organization that has a valid need for these items. In a similar manner, agents working in open societies in support of insurgent movements may openly purchase arms, ammunition, foodstuffs, medical supplies, and other needed items, normally through some front organization or nongovernmental group. The supplies are then surreptitiously transported to insurgent agents within the subject country through smuggling operations—by land, sea, or air.

In World War II Poland, the Home Army bought large quantities of artificial fertilizer from two German-controlled factories at Chorzow and Moscice through agricultural cooperatives and individual farmers. From this fertilizer, the underground extracted saltpeter for use in explosives. Conversely, in 2011, problems arose from a large flow of fertilizer from Pakistan into Afghanistan. This flow fed into improvised explosive device (IED) production facilities in Afghanistan, creating significant bilateral tension between the United States and Pakistan and leading to proposed measures to alter the composition of the fertilizer or to better track the sale and final disposition of the fertilizer.

BATTLEFIELD RECOVERY

Since ancient times, irregular forces have relied on battlefield recovery in order to supply their soldiers with arms and ammunition. Guerrilla warfare most often features irregular forces performing ambushes and raids against isolated conventional forces and then rapidly withdrawing. But the purpose behind such maneuvers often includes securing the battlefield long enough to confiscate whatever arms, ammunition, and other supplies the defeated forces left behind. Indeed, history provides many examples of underequipped armies sending men into battle without weapons for the purpose of arming them from the defeated opponents. Modern insurgencies likewise seek to obtain vehicles, communications equipment, food, medical supplies, and other items from their clashes with government military and police forces.

Thefts

Secret Confiscations

Supplies may be removed secretly from plants and warehouses by workers who are sympathetic to the insurgency. Italian workers during

World War II were able to supply the underground with some radios that were pilfered from stocks in their factories. The risk in this method was great, however, because inventories were taken regularly, and, perhaps more significantly, such confiscations could not be counted on to produce a steady supply of goods. The problem of inventory checks can be avoided, however, if office clerks are able to account for losses by forging orders and invoices, altering bookkeeping records, etc. This was done by Polish workers in two large pharmaceutical plants in Warsaw to cover the transfer of 5,400 kilograms of urotropine to the Home Army for use in explosives. Recent internal confiscations of authentic South African passports and their reappearance on captured or killed Islamic insurgents in East Africa have raised concerns in Europe and the United States and led to new visa requirements for visitors from South Africa.

Raid

Among the earliest activities of a growing insurgency are raids to acquire supplies and equipment. The Fuerzas Armadas Revolucionarias de Colombia (Revolutionary Armed Forces of Colombia, or FARC) in Colombia began by raiding farms to obtain foodstuffs and later began attacking small military outposts or patrols for the primary purpose of taking weapons, clothing, communications equipment, and ammunition.

Raids are often made on warehouses or other storage centers. In France during World War II, the manager of one warehouse was awakened by twelve masked resistance members who forced him to hand over his keys. There were trucks in the courtyard and 200 men ready to load them. A total of 38 tons of coats, sweaters, shoes, radios, and typewriters were taken. Many such raids were carried out in France after resistance workers established "understandings" with sympathetic warehouse employees.

After the Japanese evacuated Vietnam, the Viet Minh rapidly moved into their former occupiers' bases to seize abandoned equipment and supplies. In some cases, the insurgents had no training in how to use what they took, but they gradually trained themselves, sometimes aided by Chinese Communists.

In El Salvador, looting from the military and even simple recovery from abandoned military, police, and civil facilities were also effective means of acquiring needed supplies. By 1983, attacks on isolated army bases had increased so much that the Frente Farabundo Martí para la Liberación Nacional (Farabundo Marti National Liberation Front, or FMLN) "guerillas could make credible claims that most of their weapons, including even mortars and other artillery pieces, came from the United States by way of captured government troops."[3]

Likewise, the Maoist Shining Path insurgency in Peru and the Islamist Boko Haram insurgency in Nigeria initially sustained themselves by raiding police stations and mining camps to collect weapons, explosives, and supplies.[4]

Manufacturing

Types of Manufactures

Undergrounds frequently engage in the manufacture of such items as mines, flamethrowers, hand grenades, incendiaries, explosives and detonators, boots, mosquito nets, waterproof ponchos, and hammocks. Rarely, however, are they able to turn out heavy equipment because of concealment problems. One exception occurred in France during the Nazi occupation, when workers in a steel mill of Clermont-Ferrand succeeded in constructing four crude tanks out of farm tractors and sheets of steel from the factory. The components were hidden separately inside the plant until they could be welded together and armed with 87-mm cannons and heavy machine guns.

The Liberation Tigers of Tamil Eelam (LTTE) in Sri Lanka significantly altered the scope of their campaign against the government when they established a maritime capability that included high-speed boats manufactured for suicide operations against the Sri Lankan navy.[5]

Rural Manufacturing

The Viet Minh achieved a degree of safety in conducting their manufacturing in rural areas under nominal French control by using small, mobile workshops that could be moved from place to place to avoid French forays. The small size and simplicity of these shops aided their mobility—ten to fifteen workers generally were involved, and frequently manpower was the only source of energy. In spite of their crudeness, these shops were a major source of such items as mines and explosives.

Later, the Viet Cong employed the same procedures in their long struggle against the government of South Vietnam and their American allies. They would establish "jungle labs" in which they would convert captured military equipment and supplies into usable weapons and explosives.

Urban Manufacturing

An underground engaged in urban manufacturing has to use other means of avoiding the enemy. The Polish Army enlisted the services of workers in legally licensed shops, especially metal shops, to manufacture small arms. Production was thus conducted more or less in the open, avoiding the difficulty of completely hiding its noise and bustle. For camouflage, arms were sometimes produced in shops that turned out similar-looking items. Hand grenades, commonly known as "*Sidelowski*" because they closely resembled the round cans of Sidel polish, were produced in the same place as the actual cans for the polish, and flamethrowers were made in a factory engaged in the manufacture of fire extinguishers.

In Palestine, the Jewish paramilitary force Haganah used the same basic technique in the 1920s and 1930s, with variations. They established their own shops in industrial sections to avoid attracting attention. These places were devoted primarily to illegal production, although legitimate items were often manufactured at the same time so that production could be switched to "civilian" orders in case of inspections. Posted lookouts were used to warn of the approach of inspectors. Each shop was restricted to the manufacture of parts, which were more easily concealed than the finished products. By bringing the components together only at a well-hidden assembly plant, the underground also avoided the possibility of a raid on a shop in which all of the skilled workers and important machines might be captured. A natural look was also maintained by having open offices, reception desks, and office books, which were subject to inspection by auditors and tax assessors. To further ensure secrecy, only a few men in the underground—those coordinating production—knew the locations and operational features of the shops. Shop workers were selected only after extensive security checks on their backgrounds; they were also encouraged to form their own social milieu and to limit contacts with outsiders in order to lessen opportunities for security leaks.

Urban manufacturing is not always restricted to shops with legal covers, however.

The Polish Home Army underground had some small shops that were completely hidden: false walls partitioned rooms and cellars and concealed the quarters of the shops. To conceal machine noise, these shops had to be constructed near places where legal goods were being manufactured. Thus, one was built near a mechanical mangle and another just above a welding shop. In addition, work that involved use of chemicals often had to be done at night so that no one would notice the special colors of smoke rising from the chimneys.

Collections from the Populace

Goods may be systematically collected from the population, although this requires a high degree of underground influence and freedom of action. In rural areas, food is often collected for guerrilla troops. This was done in Greece during World War II. The Ethniko Apeleftherotiko Metapo (National Liberation Front, or EAM), through its "Guerrilla Commissariat," supported guerrillas through the levy of regular tithes of foodstuffs from the peasants whom it effectively controlled. In addition to these tithes, for which no payment was made, other foodstuffs were purchased at a scale of prices set by the underground.

To avoid being considered "bandit" organizations, undergrounds often make it a practice to give at least nominal payments or IOUs for goods requisitioned from peasants or other persons of modest means. Ernesto "Che" Guevara of the Cuban *Movimiento 26 de Julio* stated that the fundamental rule is to always pay for any goods taken from a friend. He also stated that when it is impossible to pay simply because of lack of money, one should always give a requisition or an IOU—something that certifies the debt.

Such consideration is not always shown, however; in Sierra Leone, RUF rebels often killed village leaders in order to send a warning to all of the villagers and then proceeded to take whatever food and provisions they desired.

Sendero Luminoso, the Maoist insurgency in Peru, was somewhat unique in that it subsisted almost entirely on support within the native population, eschewing requests for external patronage. The bulk of financial and logistical support for the highly secretive organization thus came from sympathizers within the largely indigenous, impoverished population, which in turn was pointedly separated from the senior leadership of the movement for security reasons. Sendero's ideology of championing the rights of their Indian constituents against the white, European-descended, imperialist government ensured loyalty and continued support.[6] The movement had to develop a unique sustainment concept to overcome the fact that the movement avoided external support. Their ability to operate freely in large sections of the Huallaga Valley simplified their internal lines of communication for logistics and re-supply. Logistical support for the insurgency was administered via a regional leader (commissar) who led a five-person committee that was charged with overall operational planning and execution for each region. Logistics support was generally provided by villagers, either voluntarily or through coercion, as well as a small cadre of trained and specialized logistics personnel who provided weapons and ammunition. Local villagers would routinely be directed to hide ammunition or other contraband items, with their compliance motivated either positively, by ideological and emotional support for the cause, or negatively, by fear of violence and even death.[7]

External Means

Support from Foreign Governments

Operational and logistical demands can sometimes overcome an insurgency's desire for self-sufficiency and independence from external influence. The New People's Army in the Philippines during the Ferdinand Marcos presidency was designed from the start to be self-sufficient. As the insurgency became more complex, however, the logistics requirements became greater. During 1971, the Communist Party of the Philippines established a permanent delegation in Beijing to coordinate support from the Chinese government.[8] The FMLN in El Salvador could not have sustained its insurgency without external support from the Soviet Union and Cuba, to include Soviet-produced man-portable air defense systems (MANPADS), such as the SA-7 and SA-14, and the post-war disarmament of over 10,000 weapons, 9,000 grenades, and 4 million rounds of ammunition, in addition to the 9,500 antipersonnel land mines laid by the insurgents.[9]

Foreign governments may combine direct support of insurgencies with clandestine measures to produce and supply nonattributable goods to client movements. The Iranian Revolutionary Guard Corps (IRGC) operating in Iraq after the 2003 American invasion used this method to provide Shiite militias with materials to produce improvised explosive devices (IEDs). This practice gives the sponsor some degree of deniability and is of particular concern when combined with the potential for supplying weapons of mass destruction.

Import Firms

An underground may use businesses engaged in foreign trade to import equipment under noncontraband labels, as occurred during Haganah activities. A textile firm, for example, might order textile machinery, and delivery would be in arms-producing machinery or arms parts. Payment to the firm would be made for goods or services supposedly received, thus keeping all financial records in good order. In 2010, a similar scenario emerged when thirteen shipping containers labeled "building supplies" were seized in the Nigerian port of Lagos and found to be packed with rocket launchers, mortars, explosives, and ammunition. Authorities remained unsure of whether the weapons—originally shipped from Iran—were destined for internal insurgent groups, such as the Movement for the Emancipation of the Niger Delta

(MEND) or Boko Haram, or whether the weapons were en route to Gaza via West Africa.

Parachute Drops

Supplies may also be obtained from a sponsoring government through parachute drops. Probably the most familiar instance of this type of operation is the drops that the French resistance received from Great Britain's Royal Air Force. Sophisticated radio liaison was necessary in order to work out the details of the drops. Such matters as agreement on drop-zone locations, the exact times of the drops, and ground-to-air recognition signals had to be coordinated in advance. After the drops, which usually took place at night, resistance personnel stored the goods in caches near the drop zone so they could leave the scene immediately and without incriminating evidence. Special liaison agents from abroad were often used to help execute these complex arrangements.

Wartime Equipment

Wartime stores of equipment sometimes provide a postwar source of supplies. For example, the Malayan Communist Party (MCP) was able to provide guerrillas after World War II with many arms cached during the war. These were arms originally received in air drops from the British for use against the Japanese. By claiming that many drops were lost, the Communists received extra drops, and only these extra arms were returned to the British authorities after the war. The rest remained in caches and were finally used during the "Emergency."

The beginning of the Kosovar uprising in 1997 was greatly facilitated by the collapse of the government in neighboring Albania that same year and the immediate, open access that the Kosovars, who were ethnic Albanians, gained to the ammunition and weapons depots in Albania.[10] In 2011, the revolution in Libya and the eventual demise of the Gaddafi regime resulted in the proliferation of advanced weaponry into the Sahel and Maghreb regions of Africa and the reported acquisition of Libyan weapons stores by Al Qaeda in the Islamic Maghreb (AQIM) insurgents.[11] Perhaps most concerning was the experience in Sierra Leone and the Niger Delta, where RUF and MEND insurgents, respectively, were found to have acquired weapons that were sold by Nigerian soldiers, to include African Union (AU) peacekeeping soldiers who were actively engaged in operations against the RUF and six members of the Nigerian security forces who sold 7,000 military assault rifles, sub-machine guns, and rocket-propelled grenades worth $1 million to Niger Delta militants.[12]

Transportation by Vehicles

It is often necessary to ship contraband by trucks, in which case a number of devices may be used to hide the cargoes and avoid arousing suspicion. Arms destined by the Haganah for caches in agricultural regions were often hidden in farming implements that were being taken to these places, while consignments to urban areas were frequently put in compressors, gas cylinders, asphalt sprayers, and other industrial pieces. During the orange season, truck cargoes were sometimes covered with layers of oranges, which would roll into any hole made while the cargo was being inspected. Illegal cargoes were also concealed by tarpaulins covered with fertilizer, preferably with a disagreeable odor. The chances were that policemen, well-dressed and polished, would not insist on a full inspection of such cargo. Another device was the use of trucks of well-known firms such as breweries, whose products were shipped everywhere and in great quantity. These usually escaped suspicion. Underground members dressed as policemen and driving motorcycles sometimes escorted heavy truckloads under the very "auspices of the law." Trucks even succeeded in joining British military convoys, often traveling hundreds of miles and passing many roadblocks with no check at all. It was necessary, of course, to make telephone calls and inform commanders of roadblocks that two or three lorries from another unit had been added.

The New People's Army in the Philippines established a complex intra-island and inter-island network tied to small boats called bancas. Bancas were extensively used for fishing and legitimate island trade, and it was almost impossible to distinguish an NPA banca from another banca. Moreover, the Philippine security forces lacked the brown water navy and patrol boat structure to effectively secure the thousands of miles of navigable waterways. During the three decades of active insurgent operations by the NPA, only limited logistics shipments were ever interdicted by Philippine security forces. Waterways were effectively conceded to the NPA.[13]

By Foot and on Animals

Because guerrilla bases are usually in remote areas that are difficult to access, the transport of supplies to guerrillas has usually not been mechanized. In German-occupied Greece, for instance, the rural "Guerrilla Commissariat" used pack animals as far as they could negotiate the mountain trails, and mountain dwellers carried the supplies the rest of the way. In Vietnam in the early 1950s, coolies were used extensively. One Viet Minh division required about 40,000 porters to supply its minimum needs. These coolies were local inhabitants organized into what was called the "auxiliary service." On level terrain, the coolies were expected to cover 15.3 miles per day (12.4 at night) carrying 55 pounds of rice or 33–44 pounds of arms. In mountainous areas, the day's march was shortened to about 9 miles (7.5 miles at night), and the load was reduced to 28.6 pounds of rice and 22–33 pounds of arms. In Sierra Leone, the RUF forced villagers to serve as "human caravans" and carry the food and possessions that the RUF stole from their village.

If several days or nights of travel are required, stopover facilities will be needed. Che Guevara recommends that "way stations" be established for this purpose in the houses of persons affiliated with the movement. According to Guevara, these houses should be known only to those directly in charge of supplies, and the inhabitants should be told as little as possible about the organization, even though they are trusted people.

STORAGE

Supplies are sometimes stored in individuals' houses. More often they are stored in centralized locations so that fewer persons are subject to capture in the event of searches. Caches are frequently located in remote areas. Members of the French resistance, for example, dug and camouflaged pits near parachute drops to store equipment until it could be moved to more convenient hiding places. The FMLN in El Salvador adopted the policy of burying weapons caches throughout the countryside, avoiding the establishment of large supply depots. In that regard, any capture or find by the El Salvadoran armed forces was not enough to significantly cripple military operations.[14] In Burma, the Karen National Liberation Army (KNLA) insurgents were found to be storing large caches of explosives in Karen refugee camps along the Thai border.[15] Remote areas are also used as hiding places for the benefit of guerrillas. The Malayan Min Yuen collected food in rural areas and delivered it to caches hidden in the jungle, where it was picked up by the guerrillas. In Vietnam, local inhabitants helped "prepare the battlefield" for the guerrillas by storing food near the scene of an impending Viet Minh attack. These stores enabled the guerrillas to travel lightly and quickly. Where supplies must be stored for longer than a couple of days, the caches have to be ventilated and insulated against dampness. Of course, the ventilators must be camouflaged. Pipes from underground Viet Minh caches were sometimes covered at the surface by bushes.

MAINTENANCE

Weapon and equipment maintenance is a crucial logistic function for conventional armies; it is even more so for insurgents. Because logistics are often desultory, interdicted, or otherwise unreliable, undergrounds must provide for effective maintenance of equipment that is on hand. The underground shares maintenance duties with the auxiliary and the armed component. Guerrillas perform routine unit-level maintenance on their equipment, and the auxiliary typically handles higher levels of maintenance (e.g., engine replacement, weapons repair). Insurgencies that enjoy control over relatively secure areas often build dedicated maintenance areas within their secure compounds and training camps.

MEDICAL

Undergrounds typically seek to establish effective medical services, both to sustain the end strength of the armed component and to extend such services to the population they are attempting to win over or govern. In impoverished areas or among populations that are alienated from effective government control, insurgents can secure local loyalties through the provision of vaccinations and other simple health care. As with the maintenance function, undergrounds will often establish hospitals and clinics within relatively secure areas, such as training camps. To obtain the requisite medical expertise, underground leaders will seek to recruit doctors, nurses, and medical technicians into the auxiliary.

ENDNOTES

1 Summer Newton, "The Provisional Irish Republican Army: 1969–1998," in *Assessing Revolutionary and Insurgent Strategies* (Laurel, MD: The Johns Hopkins University Applied Physics Laboratory, 2009), 64–66.

2 Defense Aerospace.com, "Arms Transfers and Trafficking in Africa," http://www.defense-aerospace.com/article-view/verbatim/16134/arms-transfers-and-trafficking-in-africa.html.

3 Ron Buikema and Matt Burger, "Farabundo Marti Frente Papa La Liberacion Nacional (FMLN)," in *Casebook on Insurgency and Revolutionary Warfare, Volume II: 1962–2009* (Laurel, MD: The Johns Hopkins University Applied Physics Laboratory, 2009).

4 Ron Buikema and Matt Burger, "Sendero Luminoso," in *Casebook on Insurgency and Revolutionary Warfare, Volume II: 1962–2009* (Laurel, MD: The Johns Hopkins University Applied Physics Laboratory, 2010), 64.

5 Maegen Nix and Shana Marshall, "Liberation Tigers of Tamil Eelam (LTTE)," in *Casebook on Insurgency and Revolutionary Warfare, Volume II: 1962–2009* (Laurel, MD: The Johns Hopkins University Applied Physics Laboratory, 2009).

[6] Ibid., 59–60.

[7] Ibid., 65.

[8] Ron Buikema and Matt Burger, "New People's Army (NPA)," in *Casebook on Insurgency and Revolutionary Warfare, Volume II: 1962–2009* (Laurel, MD: The Johns Hopkins University Applied Physics Laboratory, 2010), 17.

[9] Buikema and Burger, "FMLN."

[10] Maegen Nix and Dru Daubon, "Kosovo Liberation Army, 1996–1999," in *Casebook on Insurgency and Revolutionary Warfare, Volume II: 1962–2009* (Laurel, MD: The Johns Hopkins University Applied Physics Laboratory, 2009).

[11] Lee Ferran and Rym Momtaz, "Al Qaeda Terror Group: We 'Benefit From' Libyan Weapons," *ABC News Online*, November 10, 2011, http://abcnews.go.com/Blotter/al-qaeda-terror-group-benefit-libya-weapons/story?id=14923795.

[12] "Nigerian Soldiers Jailed for Life for Arms Sales to Rebels," *Voice of America News*, November 19, 2008, http://www.voanews.com/english/news/a-13-2008-11-19-voa34-66735022.html.

[13] Buikema and Burger, "New People's Army," 17.

[14] Buikema and Burger, "FMLN."

[15] Ron Buikema and Matt Burger, "Karen National Liberation Army (KNLA), Burma," in *Casebook on Insurgency and Revolutionary Warfare, Volume II: 1962–2009* (Laurel, MD: The Johns Hopkins University Applied Physics Laboratory, 2009).

CHAPTER 6.

TRAINING

CHAPTER CONTENTS

Robert Leonhard

INTRODUCTION:
CHARACTERISTICS OF TRAINING

The training of insurgents and terrorists necessitates the combining of hard skills, political or religious indoctrination, and psychological preparation. This last category is examined fully in the companion to this work, the second edition of *Human Factors Considerations of Undergrounds in Insurgencies*, and so will not be emphasized in this chapter. Nevertheless, radicalization and moral disengagement is a key feature of insurgent training and one of the characteristics that makes the process so different from the training of regular military forces.

The focus of this chapter is the function of training that underlies the development of underground, auxiliary, and guerrilla forces. It is at once obvious why insurgencies require training: guerrillas, saboteurs, terrorists, administrative specialists, messengers, and other operatives can accomplish their tasks only if they are competent in the use of weapons, explosives, communications equipment, and so forth. But insurgent training is driven by purposes beyond mere skill building. Trainers also use the training process to enhance the movement's security, provide propaganda and indoctrination, and select and develop future leaders.

A new member's training after joining an insurgent movement is typically heavy in propaganda, historical justification, and ideology. In this respect, the underground function of training differs significantly from the training of a conscripted or professional soldier. In the latter case, a recruit comes from a society and a schooling background that likely look favorably upon (or at least are not hostile toward) military service. Normal childhood development provides ideological preparation for later service, and military training institutions do not have to spend much time reiterating historical justification and ideological grounds for service to the recruits. This is not the case in insurgent training. Instead, leaders create a training environment rich in ideology so as to immerse the recruit in a new set of corporate values designed to supplant his former loyalty to society or the government.

During this period of initial training, insurgent leaders evaluate new recruits for political (or religious) reliability. Thus, initial entrance training serves the dual functions of skill building and gate keeping. Only demonstrated ability as well as fidelity to the cause will lead to more advanced training for the recruit. In addition, leaders evaluate the prospective new member's personal habits and demeanor, sometimes rejecting those considered too immature, too given to drink, or too talkative as security risks.

Because insurgency training must be conducted clandestinely, it typically takes longer than similar training in a conventional military force. Training tends to occur sporadically in intervals of short duration. Likewise, insurgency training tends to focus on training of individuals or very small units, rather than on the collective training of larger units. This is in part because of the difficulty of maintaining secrecy when training large groups, and also because insurgent military operations feature individual and small group actions.

An insurgent movement grows through experience in combat against the government. As a result, another characteristic of insurgency training is the routine application of lessons learned in battle. Professional military establishments also make use of experience to modify and update training, but the process is less formalized, more immediate, and more pronounced within insurgencies because there is typically less room for error and failure.

Conventional military forces—especially the most successful ones—focus on sustained operations. That is, planners, leaders, and soldiers are trained to think beyond the preliminary stages of a battle, campaign, or war and instead to focus on the sustained application of combat power all the way through to mission completion—often described as the destruction of enemy forces or the control of terrain. Insurgencies, conversely, tend to focus instead on the initiation of armed conflict, because the operational assumption is that they will have a short time interval during which to attack and that the attack will be followed by withdrawal, dispersal, and hiding from pursuing government forces. Hence, training often concentrates on achieving an effective first shot.

These characteristics of insurgent training—multiple purposes, strong ideological component, gate keeping, clandestine environment, focus on individual training, inculcation of recent combat experience, and emphasis on the initiation of combat—drive insurgent leaders to develop efficient ways and means for training their members. Two of the most frequently used methods are the use of training camps and online training.

TRAINING CAMPS

Training camps are the epicenter of terrorist and insurgent training throughout the world. As described in numerous first-hand accounts, the camps provide an environment of isolation, focused skill building, and indoctrination that can mass produce deadly and committed warriors.

Al Qaeda maintained secret training camps in Afghanistan and Bosnia that were later exposed and destroyed. But these well-known examples are in fact typical of similar facilities in every corner of the globe. The Bekaa Valley in Lebanon has long hosted training camps for both Hizbollah and the Palestine Liberation Organization (PLO). Libya, Algeria, and Tunisia hosted training camps for terrorists and insurgents throughout the late twentieth century. Jemaah Islamiyah (JI) ran several effective training facilities that focused on weapons and explosives training for Indonesian terrorists. Likewise, both the Moro Islamic Liberation Front (MILF) and JI built training camps in Mindanao in the Philippines. The Tamil Tigers of Sri Lanka based several training sites in the remote north of their country.

Figure 6-1. Photograph of the Garmabak Ghar Terrorist Training Camp, Afghanistan. Used by Secretary of Defense William S. Cohen and Gen. Henry H. Shelton, U.S. Army, chairman, Joint Chiefs of Staff, to brief reporters in the Pentagon on the U.S. military strike on a chemical weapons plant in Sudan and terrorist training camps in Afghanistan on August 20, 1998. (Released; retrieved from http://www.fas.org/irp/imint/011009-D-6570C-004.jpg.)

The Colombian insurgency Fuerzas Armadas Revolucionarias de Colombia (Revolutionary Armed Forces of Colombia, or FARC) used training camps extensively, capitalizing on the inaccessibility of their terrain as security against government discovery of their facilities. As mirrored within other insurgent training regimes, FARC's included both hard skills and ideological topics.

> [I]n 1982, the FARC established a guerrilla training center called la Escuela de Cadetes (the Hernando González School of Cadets). Military topics included "mobile guerrilla warfare, military psychology, urban guerrilla warfare, communications, first aid, cartography, artillery, and related topics." New recruits trained locally through a four-month course that included both military and ideological disciplines and the FARC's organization and code of justice.[1]

The Viet Cong's use of clandestine and isolated training camps illustrates the multipurpose nature of training:

> Viet Cong recruits normally received training at a location outside of their villages, where they also were subjected to extensive political indoctrination. The training was usually conventional (i.e., weapons use, marching, small-units tactics, etc.). Recruits who displayed the most potential received additional advanced training (as well as indoctrination) and were assigned to the main force units; the rest were used for guerrilla combat mission and other (part-time) assignments . . . Main Force Viet Cong units, or the "the fighting arm of the organization," were made up of men who tended to be extremely motivated. They were directed by cadre and were full-time, uniformed soldiers. Typically employed to launch large-scale offensives over a geographically wide area, Main Force units had an average experience of over two decades in insurgency warfare and many had fought against the French in the First Indochina War—some having even fought in the resistance movement against the Japanese in WWII. The fighters of the Main Force units were normally trained at recruiting camps separate from those of the guerrilla units.[2]

Insurgents have built training camps in out-of-the-way locations in Ireland, France, Peru, and the United States to train, indoctrinate, and radicalize potential terrorists and warriors for the cause. The Aryan Nations Church, for example, maintained a twenty-acre guarded camp in Northern Idaho where Christian Identity members received weapons and tactics training. Another Christian group, The Covenant, The Sword, and the Arm of the Lord, constructed a 224-acre base called Zarephath-Horeb on the Missouri–Arkansas border, where they collected a huge cache of weapons and ammunition.

A recent study of captured student notebooks from arrested terrorists in Uzbekistan provides insights into the nature of training in clandestine camps. The students, who trained throughout the mid-1990s, received instruction in map reading, small arms, demolitions, and poisons, along with ideological instruction in matters of jihad. Some were also instructed in Arabic and Russian, and there was some religious instruction as well, although the captured notes indicate that both the teachers and students lacked familiarity with the fundamentals of Islam. Detailed analysis of the notebooks indicates rigorous instruction in matters of targeting enemy vehicles (both air and ground) under various conditions. As described above, this emphasis confirms the insurgent focus on making the first shot count.[3]

What religious instruction the students received revealed poor scholarship and a focus on the nature and justification of jihad. The notes indicate a crude theme of anti-Semitism and xenophobia mixed with a sustained paranoia that Islam was threatened from without and within. This threat served as the justification to break with societal norms, peaceful religious instruction (the teachers of which were branded heretics), and respect for the government. In the place of these old loyalties, students were to prepare themselves for killing the enemies of Allah. These enemies included Israeli, Russian, and American visitors, along with Christians who attempt to proselytize or insult Islam. The students were to destroy any foreign business interests and any imported goods they found.[4]

After the Israeli invasion of Lebanon in 1982, elements of Iran's Revolutionary Guards Corps (IRGC) moved into the Bekaa Valley and established the infrastructure for the formation of what would become Hizbollah. The Syrian-held town of Baalbek was the original site for their fledgling operation. Reinforced by clerics and agents from Iran's Ministry of Foreign Affairs and Ministry of Intelligence and Security, the Guards conducted military training and religious indoctrination for the embryonic insurgency. In late 1982, the Iranian-led cadres fanned out to outlying villages throughout the valley to recruit and indoctrinate potential members for the emerging insurgency movement. By 1984, the IRGC was operating six training camps in the Bekaa Valley and providing salaries, medical benefits, and free education for fighters and their families. The IRGC's main focus remained on acting as a conduit to Hizbollah for Iranian arms, money, and advanced training in guerrilla and terrorist operations.

As Hizbollah developed through the 1990s, the organization became adept at hiding training camp activities. In part, this ability to remain clandestine was due to the degree to which Hizbollah became thoroughly integrated into Lebanese society. With hundreds of thousands of members and sympathizers, the organization's militant activities could easily blend in with its social and political operations. Likewise, the post-civil war decision to legitimize Hizbollah's retention of arms and its self-appointed mission as Lebanon's main military force to oppose any future Israeli incursions blurred the distinction between terrorist training and the quasi-legal training of soldiers.

Hizbollah's selection program for new recruits took about two years, during which time the young soldiers were thoroughly investigated for political and religious reliability, as well as for aptitude for special services. Once ensconced in a training camp, recruits received training and instruction in weapons, explosives, ambush techniques, infiltration, intelligence, and myriad other subjects, including psychological warfare. Hizbollah and IRGC agents paid close attention to actual combat operations, sifting through lessons learned in order to strengthen subsequent attacks.[5]

Al Qaeda's training regime, like that of many other terrorist groups, puts a premium on religious indoctrination, which Al Qaeda considers to be infinitely more important than the development of military skills. The exact ideology taught, however, differs from camp to camp, based on the cultural context of each particular insurgency. In camps supporting ethnic-based insurgencies, for example, the ideology focuses more on the history and mythology of the subject's ethnicity and how it has been wronged. Other camps that develop jihadists for the restoration of the Caliphate emphasize jihad as a religious duty.

In addition to ideological preparation, the camps have become increasingly sophisticated in the training of hard skills. Weapons and guerrilla warfare courses are followed by training in advanced techniques of assassination, surveillance/counter-surveillance, document forging, suicide operations, and heavy weapons. Since the 1980s, when the United States provided support (including weapons and training) to the mujahidin in Afghanistan, Al Qaeda experts have relied heavily on American and British doctrinal sources—field manuals, training circulars, etc.

Training camps established in permissive or semi-permissive countries can evolve into highly effective institutions for training terrorists. Al Qaeda's camp at Darunta, near Jalalabad, Afghanistan, was one such location and typifies the smaller facilities. Only about a quarter-square mile in size, it consisted of a tunnel complex, four sub-camps, each with a different purpose, and a defensive system of trenches and outposts. The entire complex was camouflaged. Within the camp, Al Qaeda ran a chemical training laboratory and a guerrilla training center. The Taliban owned a part of Darunta, and Pakistani terrorists operating in Kashmir ran the fourth sub-camp. Investigation of the abandoned site revealed extensive training and indoctrination materials—many in English, and much of it downloaded from the Internet. Darunta produced many terrorists, including Raed Hijazi (alleged to have planned attacks in Jordan) and Ahmed Ressam (convicted for plotting to blow up Los Angeles International Airport during the millennium celebrations).

The Zawar Kili complex near Khowst in Paktia province, Afghanistan, was a huge installation featuring a base camp, tunnels, and a cave system that was improved to reach deep into Sodyaki Ghar mountain. Al Qaeda used the location as a logistics base, command and control center, and training base. It boasted a hotel, repair facilities, arms depots, a medical clinic, and a mosque. Generators provided the requisite power for the extensive compound. After the Africa embassy bombings of 1998, the United States targeted the complex with cruise missiles.[6]

TRAINING ONLINE

The evolution of the Internet, personal computers, portable digital devices, and wireless communications has revolutionized many aspects of society and conflict. The underground function of training insurgents, resisters, and terrorists is one such dimension. Prior to the advent of the Internet, access to information sources on weapons, explosives, tactics, etc., was fairly limited. Today, with the click of a mouse button, anyone can download a manual that instructs him or her how to carry out illegal attacks and other operations.

Online resources for insurgent and terrorist training include both motivational and operational information. Most often, these two categories are combined into a single document because the psychological preparation of a student is deemed to be equally important as his or her development of hard skills. Motivational information comprises psychological, sociological, political, and/or religious components. The motivational approach often includes an explanation of "the right way"—i.e., how God, the Constitution, historical predecessors, or some other past or mythical phenomenon represented the perfect moral, spiritual, or political condition. In contrast to this right way, the motivational literature will then describe the present condition that requires correction. The appeal is for the audience member—alone or in concert with others—to become the agent of correction.

Hizbollah, under the direction of its senior leaders, developed a video game called "Special Force," in which players experience a simulated operation against Israeli soldiers based on real-life events. The game was released in 2003 and allowed players to conduct "target practice" against Israeli political leaders. Thousands of copies of the game were sold in the Middle East, in the United States, and throughout the world. Through the publication of a game, Hizbollah was able to export both its ideology and a form of skill building that would prepare youngsters to one day assume the role of jihadist. Hizbollah copied this training technique from American supremacist groups that offer, on their websites, similar games engendering racial hatred. These games allow players to kill Jews, black people, or other targeted groups in "first-person shooter" formats.[7]

Figure 6-2. Screenshot from "Special Force." Retrieved from http://en.wikipedia.org/wiki/File:Spf9.jpg.

Examples of online resources for insurgent and terrorist training include *The Terrorists' Handbook* (http://www.capricorn.org/~akira/home/terror.html); *How to Make Bombs, Book Two* (http://www.filestube.com/e/explosives+how+to+make+bombs+book+2); the Irish Republican Army's *Green Book* (http://www.quikmaneuvers.com/ira_green_book.html); *The Turner Diaries* (http://www.victoryforever.com/turnerdiaries.pdf) and the *Field Manual of the Free Militia* (http://www.rickross.com/reference/militia/militia10.html), used by American Christian militias; Ḥarakat al-Muqāwamah al-'Islāmiyyah's (Islamic Resistance Movement, or HAMAS) *Mujahideen Poisons Handbook* (http://tkolb.net/FireReports/PoisonsHandbook.pdf); and *The Al-Qaeda Handbook* (http://www.justice.gov/ag/manualpart1_1.pdf).

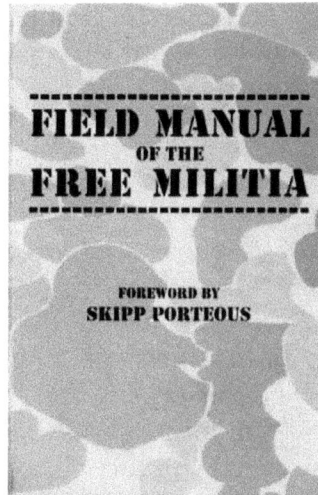

Figure 6-3. *Field Manual of the Free Militia,* **offered for sale on Amazon.com**: http://www.amazon.com/Field-Manual-Free-Militia/dp/B000AV32JA.

The widespread availability of such resources both facilitates training and serves as a recruitment and self-radicalization engine throughout the world. Timothy McVeigh, the American terrorist whose bombing of a federal building in Oklahoma City killed 168 people, was strongly influenced by *The Turner Diaries* and similar literature. Radical literature coupled with detailed instructions on the making and use of weapons has become easily accessible to all and created a new dimension in the fight against terrorism. From the insurgent's point of view, however, these developments are a godsend.

The most salient expression of online insurgent training is the ever-growing community of hackers. Thousands of new websites emerge annually that offer instruction and tools for hackers. While such sites offer little motivational information, they are replete with operational know-how, tips, tricks, and "best practices." Visitors to such sites can learn detailed techniques for conducting denial of service attacks, stealing passwords, overloading websites, and probing networks for vulnerabilities. They can also download tools for encryption, programming, and data manipulation to facilitate their efforts. Islamic jihadist groups all over the world have devoted resources to encouraging their followers to conduct cyberterror attacks and providing training and tools to assist them.

CONDUCT OF TRAINING

Training is central to the success of an insurgency. The most successful and long-lived movements treat this key underground function as a major area of concern and focus for leadership. The experience of the Provisional Irish Republican Army is instructive as an example of how an insurgent organization plans, organizes, and conducts training.

The Provisional Irish Republican Army (PIRA)

The Provisional Irish Republican Army organized a training department under its general headquarters with the responsibility to maintain all training resources and facilities. They conducted training in three areas: new recruit training, operational skills training, and intelligence/counterintelligence/security training. During new recruit training, the emphasis was on motivational information—i.e., what it means to be a Republican and the history of Irish resistance against the British occupation. The latter two phases of training focused on the necessary hard skills to conduct operations and to protect the security of the organization.

New recruits were required to attend training sessions about once per week during their first three months in the organization. The sessions included lectures and discussions about member duties, the history of the organization, the rules concerning military engagement, and how to resist interrogation. During this period of initial training, the recruit was also evaluated as to his potential for service and his risk to the security of the organization.

The PIRA learned to emphasize rigorous training and instruction in hard skills—weapons, explosives, urban and rural tactics, etc. The cost for training deficiencies in these areas is primarily ineffectiveness. Unsophisticated attacks by impulsive and unskilled youths merely invited arrests, interrogations, and political failure. Likewise, inexperience with weapons and explosives caused numerous deaths among the insurgents through fratricide and premature explosions.

Eventually, the PIRA developed a military training program that put recruits into covert training camps where they learned to shoot and maintain weapons, employ demolitions, and other basic skills. Due in part to the requirement for secrecy, the average PIRA insurgent's training took about six months. The difficult security environment caused leaders to emphasize individual training over collective training.

Before the advent of the Internet, it was difficult for insurgent leaders to get access to training resources. PIRA operatives solved this deficiency by recruiting former military members, obtaining printed military manuals, and in some cases commissioning members to pursue education opportunities that they could use in later insurgent activities. In the 1970s, PIRA leaders had to devote resources to producing written materials to support training. This resulted in, among other products, the infamous *Green Book*, which included both ideological and operational information for aspiring members.

These efforts paid off in better operational performance. Better-trained insurgents began to operate collectively instead of individually, giving them the ability to stand their ground in pitched skirmishes with government forces. Marksmanship improved, and British casualties consequently grew.

Beyond the general training of recruits in weapons and tactics, PIRA leaders also sought to improve performance in bomb making, sniping, logistics, and intelligence. Specialists in these areas would occasionally come together to receive training and pass on lessons learned with the intent of improving safety, security, and performance in battle. The ever-present danger of compromise made these meetings and collective training in general a risky operation. If a bomb maker, for example, was interrogated or turned disloyal, he might divulge the identities of other bomb makers and disrupt the entire insurgency.

Clandestine training requirements do not facilitate live-fire weapons training. Weapons ranges—especially those designed to handle mortars, explosives, and other large weapons—tend to be noisy and hard to conceal. The PIRA used remote locations throughout Ireland, including abandoned farm houses, unused beaches, and woods. In one case, they used a beach for mortar fire using dummy (i.e., non-explosive) shells. In other cases, they positioned their live-fire ranges near army training facilities so that their gunfire noise would not attract attention. Likewise, recruits were often kept ignorant of the exact location of the camps where they trained in an attempt to prevent the authorities from discovering and shutting them down. The PIRA also turned to other sympathetic groups abroad for training support. They had relationships with Ethniki Organosis Kyprion Agoniston (National Organization of Cypriot Fighters, or EOKA) (Greek Cypriot terrorists), the Basque group Euskadi Ta Askatasuna (Basque Homeland and Freedom, or ETA), Fatah, the Popular Front for the Liberation of Palestine (PFLP), and the PLO in the Middle East. PIRA agents trained abroad with these groups, including in training camps in Libya.

CONCLUSION

Training remains one of the core functions of an insurgent underground. Through the training regime, underground leaders select, evaluate, and develop recruits to populate the underground, guerrilla, and auxiliary forces. Recent developments, including the expansion of the Internet, the privatization of security operations, the end of the

Cold War, and the advent of globalized jihad, have combined to facilitate effective insurgent training to a degree previously unheard of. The insurgent leader's traditional problem of lacking access to training materials has been supplanted by almost unfettered access to advanced information on a wide spectrum of useful hard skills: weapons, explosives, guerrilla tactics, surveillance, communications, and so on.

Insurgent training differs considerably from the training of conventional forces. The primary difference is the emphasis on ideological preparation. Because leaders must replace the recruit's previous loyalty to the government with an allegiance to a radical ideology, training includes immersion in propaganda, whether religious or political. A recruit's mastery of and devotion to this ideology determines his or her potential for advancement within the movement.

Insurgent training is also characterized by its adaptability to the conflict environment. In most cases, training operations must take place clandestinely, which in turn affects the content and conduct of training. Clandestine training most often aims at training individual recruits and very small units, because dealing with larger units would compromise security. Likewise, training tends to aim at optimizing the opening minutes of an insurgent attack, rather than sustained land combat operations. Finally, clandestine training tends to take longer and unfold through separate episodes. Conversely, training in permissive or semi-permissive environments, such as those enjoyed by Al Qaeda in Afghanistan, allows for more intensive and continuous training.

Another key difference between the training of insurgents and conventional forces is the gate-keeping role. In the former, recruits are constantly evaluated for reliability, aptitude, maturity, ideological fervor, and military competence. Only those who pass the scrutiny of underground leaders are admitted to more advanced training and given access to the movement.

Finally, insurgents put a premium on the rapid assimilation of lessons learned. Battles, acts of sabotage, terrorist attacks, and other operations—whether they succeeded or failed—are scrutinized for mistakes and best practices. These lessons learned are rapidly converted into new training techniques with a view to improving the next attacks before security forces can adapt.

ENDNOTES

[1] Steven Phillips, "Fuerzas Armadas Revolucionarias De Colombia – FARC," in *Assessing Revolutionary and Insurgent Strategies—Tier I Case Studies*, ed. Chuck Crossett (Laurel, MD: The Johns Hopkins University Applied Physics Laboratory, 2010), 23.

2   Bryan Gervais and Jerome Conley, "Viet Cong: National Liberation Front for South Vietnam," in *Assessing Revolutionary and Insurgent Strategies*, ed. Chuck Crossett (Laurel, MD: The Johns Hopkins University Applied Physics Laboratory, 2010), 18–19.

3   Martha Brill Olcott and Bakhtiyar Babajanov, "Teaching New Terrorist Recruits: A Review of Training Manuals from the Uzbekistan Mujahideen," in *The Making of a Terrorist: Recruitment, Training, and Root Causes*, ed. James J. F. Forest, vol. 2 (Westport, CT: Praeger Security International, 2005), 136–151.

4   Ibid.

5   Magnus Ranstorp, "The Hizballah Training Camps of Lebanon," in *The Making of a Terrorist: Recruitment, Training, and Root Causes*, ed. James J. F. Forest, vol. 2 (Westport, CT: Praeger Security International, 2005), 243–262.

6   Rohan Gunaratna and Arabinda Acharya, "The Terrorist Training Camps of al Qaeda," in *The Making of a Terrorist: Recruitment, Training, and Root Causes*, ed. James J. F. Forest, vol. 2 (Westport, CT: Praeger Security International, 2005), 172–193.

7   James J. F. Forest, "Teaching Terrorism: Dimensions of Information and Technology," in *The Making of a Terrorist: Recruitment, Training, and Root Causes*, ed. James J. F. Forest, vol. 2 (Westport, CT: Praeger Security International, 2005), 84–97.

CHAPTER 7.

COMMUNICATIONS

CHAPTER CONTENTS

Robert Leonhard

INTRODUCTION

Communications refers to the exchanging of information, orders, intelligence, requests for assistance, and other messaging that occurs within any organization. Underground operations, including coordination with the armed, auxiliary, or public components, cannot occur in the absence of a communications system. Underground communications are characterized by their (usually) clandestine nature due to the need for security. As with much of its functionality, the underground must balance its need for communications with the risk of exposing personnel, plans, and facilities if messages are intercepted.

The underground's communications system evolves along with the insurgency or resistance itself. At first, communications typically lack rigorous discipline and uniformity. As the movement grows and operations become more complex, successful undergrounds develop secure, redundant systems for communicating. To avoid presenting an easily defeated pattern, the underground will build a system of communications using various and diverse means: face-to-face meetings, couriers, mail, dead drops, radio, cell phones, the Internet, and social media, to name a few. When the movement grows both in numbers and effectiveness so that the government's countermeasures increase in sophistication, the underground must respond with the use of codes, frequency-hopping radios, spread spectrum, and other measures to secure their communications. The rapidly accelerating rate of technological change in the twenty-first century creates a sustained conflict between insurgents and the government over the battleground of the electromagnetic spectrum. Within that context, various communications tools and media may favor the insurgent or the counterinsurgent.

NONTECHNICAL COMMUNICATIONS

While some undergrounds operating in modern countries increasingly rely on technical means for communications, all such movements employ nontechnical means as well. Such methods have the advantage of leaving no detectable electronic signature vulnerable to interception, but they can also risk the physical security of the messengers, both leaders and couriers.

Nontechnical means of communications include face-to-face meetings, the use of couriers, the mail system, and so-called "dead drops." A dead drop refers to the use of a secret location where messages, money, or supplies are deposited for later pickup by another agent. The advantage of a dead drop is that underground personnel (or the auxiliary or guerrilla they are communicating with) do not have to physically meet. Typically, an underground operative will drop the item at a pre-arranged location and then display some physical signal that only the other contact or his/her organization would recognize. The signal alerts the recipients to the fact that an item is ready for pickup at the dead drop site.

Nontechnical means of communications can be among the most secure methods as long as the personnel involved are not at risk of government surveillance. Successful undergrounds carefully manage the use of these methods to ensure discipline; proper security precautions; variation in drop sites, meeting sites, or mail addresses; and, when necessary, counter-surveillance.

Critical messages, such as emergency warnings to other units, must get through with speed and certainty. Undergrounds therefore use parallel communication nets—that is, an important message is sent by two routes in order to ensure delivery. For less vital messages, the underground uses a backup system: the message is sent, and without immediate confirmation that the message arrived, another message is sent through a different channel. In all clandestine operations, acknowledgment of the receipt of a message is crucial. If positive confirmation is not received, or if there is considerable delay, the underground must assume for security reasons that the message has been intercepted and that the network may be compromised. In that event, the fail-safe principle is applied: if a message is believed to have been intercepted, members who are most susceptible to compromise either prepare cover activities or proceed to some safe house and await events to determine whether the ring has been compromised and whether it is necessary for them to use an escape and evasion net.

Parallel communications are important because it takes considerable time to set up a communication net in a hostile environment. Every effort is made to have alternative means of communication available in case there should be a breakdown in any one channel. In Malaya, the Communist Party of the 1940s and 1950s was forced to rely entirely on runners to deliver messages, and this constituted one of its major handicaps. A message may go to one address, be picked up by a courier and delivered to another place, and thence be delivered by another courier to the recipient. In this way, the message moves through a chain-like system of mail drops. Obviously, for routine messages, it is advisable to use mail drops or couriers because they ensure the security of the various units. Emergencies may require that the message be sent directly by courier or by technical means.

TECHNICAL COMMUNICATIONS

Undergrounds, as well as their partners within the auxiliary and guerrilla forces, frequently make use of radios of all kind—from ultra-high-frequency radios to ham radios to citizens' band radios. Although radio usage can be fairly easily jammed, intercepted, or exploited for direction finding, radios are cheap, expendable, and do not require traceable accounts like telephones. Insurgencies and resistance movements operating in permissive or semi-permissive environments, or in areas in which opposing forces lack effective signal intercept/jamming equipment, can make very effective use of radios to coordinate operations.

The emergence of cell phone technology throughout the first decade of the twenty-first century has added a new dimension to communications among resisters and insurgents. From the start, some analysts have pointed to the undeniable fact that the cell phone can assist clandestine operations and exacerbates insurgencies.[1] There is, however, evidence in the opposite direction as well, demonstrating that counterinsurgents can benefit equally from the use of the cell phone and that an insurgent's use of the device frequently leads to intercepts, monitoring, and key intelligence finds.[2]

Iraqi insurgents operating in the aftermath of the American invasion (2003–2011) made frequent use of cell phones in planning and coordinating attacks against coalition forces and other targets. Cellular devices were used not just for communications but also as the triggering devices for improvised explosive devices. The irony was that the United States sponsored the construction of numerous cell phone towers throughout the country, which were then used by insurgents against American troops and officials. At the same time, cell phone intercepts also serve the needs of the counterinsurgent: cell phones played a role in the American targeting of Abu Musab Al-Zarqawi and later Osama bin Laden. With widespread proliferation of cell phones, the population can anonymously phone in tips and alerts when they observe suspicious behavior.

Resistance and insurgent movements that operate in urban environments are more likely to be able to exploit cell phone technology than those in rural environments like Afghanistan. Indeed, the Taliban were known to shut down cell phone towers at night in order to prevent their use in facilitating American operations. Urban insurgents in Iraq not only refrained from destroying or shutting down cell phone towers but also, in some cases, threatened or coerced commercial companies to maintain the towers because they were so useful in insurgent operations. Thus, both the insurgent and counterinsurgent must analyze the situation to determine the advantages and disadvantages of cell phone systems. Like the weather, cellular communications may favor either side in the struggle.

With about one-third of the world's population able to use the Internet, computer-based communications methods have begun to impact both insurgency and counterinsurgency in the modern world. The Internet offers the insurgent the ability to use e-mail, chat rooms, blogs, and other forums both for propaganda and for direct messaging. Operatives can capture digital video imagery, upload it within minutes, and broadcast it around the world almost instantaneously. This new capability has completely undermined the traditional control of information that tyrannical governments have exploited in the past. And because Internet usage is either free or fairly cheap, underground operatives can use it instead of having to pay for printing materials or delivering messages through physical means.[3]

The Orange Revolution in Ukraine that unfolded from November 2004 through January of the following year featured perhaps the first widespread use of the Internet to help foster revolution. Web postings, combined with the use of cell phones, bolstered ever-growing crowds protesting the rigged elections in Kiev. By using technology that could not be easily interdicted or controlled by authorities, the crowds organized demonstrations, sit-ins, and strikes to compel the government to annul the suspicious election results. The outrage felt in the streets of Kiev found an international audience through the new technology, and the result was the deposing of a tyrant in favor of Victor Yushchenko.

Four years later, popular resistance in Moldova capitalized on Twitter, the Internet-based social messaging service. Angered by perceptions of fraud in parliamentary elections, citizens erupted in demonstrations and rioting in April 2009. They used Twitter to incite unrest, provide updates on protesters' actions and the government's reactions, and appeal to the international community for help. The protesters used "hashtags" to group messages under popular headlines, such as "#pman," which stood for "Piata Marii Adunari Nationale," the name of the biggest square in Moldova's capital of Chisinau. The use of the messaging service was so influential in the unrest that the incident became associated with the so-called "Twitter Revolution."[4]

Later that year in June, the Green Movement in Iran—an abortive attempt to overthrow the government of President Mahmoud Ahmadinejad—made effective use of Twitter and YouTube to garner international attention. The amateur filming of the shooting death of a young woman named Neda Agha-Soltan appeared on Facebook and YouTube and incited outrage both within Iran and around the world. Protesters even conducted denial-of-service attacks against government websites, prompting the Iranian government to shut down Internet access. As the massive unrest continued, the government also shut down or limited cell phone usage and reinstated Internet access with low bandwidth to try to prevent video footage from being used. The conflict saw both sides very determined to control cyberspace for their own purposes, and both the government and the protesters demonstrated growing technological sophistication and innovation in their attempts to thwart each other.

The "Arab Spring" of 2011 likewise featured widespread and effective use of Internet-based social media to bolster insurgency. Twitter, Facebook, YouTube, and other services were used to publish protesters' propaganda, pictures, and video. Through such methods, leaders could recruit sympathetic sectors of the population to participate in demonstrations and strikes. Because the younger generation is more likely to use social media, they responded more readily than the middle-aged and elderly.

Mohammed Nabbous, a Libyan businessman and technologist, established an Internet television station that he named "Libya Alhurra" (Free Libya) in Benghazi in February 2011. Able to thwart government attempts to shut down his broadcasts, Nabbous found a worldwide audience and contributed both to the growing insurgency within Libya and to international outrage and eventual action from NATO. He was eventually shot and killed in a gun battle, but his efforts helped propel the anti-Gaddafi forces to victory over the tyrannical regime.

Likewise, social media played a significant role in the unrest in Egypt. Google executive Wael Ghonim used his expertise in Internet marketing to create an insurgency group under the banner "We Are All Khaled Said"—a reference to a recent victim of torture and execution by the Egyptian police. He and others used Internet resources and cell phones to incite "flash mobs" and other demonstrations that attracted international attention and ate away at government legitimacy. The combined power and ubiquity of the insurgents' message through global media influenced political leaders in the West, including in the United States, to pressure President Hosni Mubarak to step down.

Insurgents and terrorists also employ the Internet in attempts to demoralize their opposition. Videos of successful ambushes or sniping often appear on jihadist websites and other venues in order to encourage fellow travelers and to demonstrate the resiliency of the resistance to the enemy. Government officials tend to ignore or downplay such propaganda, but popular reaction is often much more pronounced. Insurgents can use such techniques to stimulate a sense of hopelessness within enemy populations in the hope that the citizens will in turn influence their governments to withdraw.[5]

Social media can provide citizens with the opportunity to partici-pate in an insurgency or revolution from relative safety, because they can enter the fray without leaving their homes. Recruitment, financing, training, propaganda, and even intelligence operations can take place in real time and attract a disproportionate audience within cyberspace. Thus, the dynamics of Internet-based social media and the cell phone have demonstrated the potential to revolutionize the business of insur-gency and counterinsurgency. The resulting technological conflict accelerates innovative workarounds such as "bouncing"—i.e., posting material on multiple servers in different locations, and the use of mes-saging technologies that defy attempts at governmental control.

Because social media is by its very nature decentralized, its useful-ness in reaching and mobilizing masses is not equaled by its utility in planning and controlling the day-to-day operations of an insurgent movement. It is easy to reach out and recruit a young enthusiast into the movement; it is much harder to direct, control, and discipline that person and make him or her useful. Classic Maoist insurgencies focus on tight control by a well-organized and powerful central group of lead-ers. But modern communications may serve to undermine and replace that approach to organization and leadership.

The Arab Spring movements in Tunisia, Libya, Egypt, and, to a lesser degree, Syria fostered a new enthusiasm for the potential of social media in part because the insurgents were young, Internet-savvy people fighting against old, traditional dictatorships. Government attempts at disinformation, particularly in Libya, Egypt, and Syria, came across as ham-handed and clumsy when compared with the skill with which the younger insurgents mastered their Internet resources. But as the tech-nology continues to grow and penetrate further into all sectors of the world population, governments will likely catch up and come to view social media as the effective tool—or weapon—that it really is.[6]

CONCLUSION

Undergrounds must employ effective communications to survive and grow an insurgency. They must communicate with members of the underground, the auxiliary, the guerrilla forces, and, when appli-cable, the public component of the movement. In addition, they must be adept at communicating their messages to the outside world and even to the government they oppose.

Over the past forty years, computers, the Internet, cell phones, and social media have combined to revolutionize communications. Insur-gency and counterinsurgency now play out in a radically different con-text than their Cold War counterparts experienced. Mastering the new media as well as the older nontechnical forms of communication is a *sine qua non* for success in the modern world. Because the pace of technological change is likely to increase, modern insurgents and counterinsurgents require new skills and mindsets with regard to the means and methods of communications.

ENDNOTES

[1] John Arquilla, David Ronfeldt, and Michele Zanini, "Networks, Netwar, and Information-Age Terrorism," in *Strategic Appraisal: The Changing Role of Information in Warfare*, ed. Zalmay M. Khalilzad and John P. White (Santa Monica, CA: RAND, 1999).

[2] Jacob N. Shapiro and Nils B. Weidmann, "Talking About Killing: Cell Phones, Collective Action, and Insurgent Violence in Iraq," September 6, 2011, https://bc.sas.upenn.edu/system/files/Shapiro_09.29.11.pdf.

[3] Sean Kennedy, "New Media: A Boon for Insurgents or Counterinsurgents?" *Small Wars Journal*, September 4, 2011, http://smallwarsjournal.com/node/11414.

[4] Evgeny Morozov, "Moldova's Twitter Revolution," *Net Effect (blog)*, April 7, 2009, http://neteffect.foreignpolicy.com/posts/2009/04/07/moldovas_twitter_revolution.

[5] Bruce Hoffman, "The Use of the Internet By Islamic Extremists" (*testimony presented to the House Permanent Select Committee on Intelligence*, May 4, 2006).

[6] Clay Shirky, "The Political Power of Social Media: Technology, the Public Sphere, and Political Change," *Foreign Policy*, January/February 2011.

CHAPTER 8.

SECURITY

CHAPTER CONTENTS

Jerome M. Conley

INTRODUCTION

Security is the primary consideration for an underground and permeates the entire movement—its organizational structure, members, and activities—and is a
prerequisite for the movement's survival.

—*The Underground* (ST 31-202, dated August 16, 1978)[1]

Through the establishment and enforcement of security measures,
an underground is able to ensure its ability to provide administrative
and operational support to the insurgency as it evolves during the early
stages of growth and activities. This chapter highlights key concepts of
security that have been adopted by previous revolutionary movements
and also provides examples of breaches in these security measures and
the resultant consequences for the associated insurgency.

FORMATION AND ORGANIZATION OF
THE UNDERGROUND

From the initial steps of an insurgency forming, the core leadership of the movement must consider basic issues of secrecy and security
for the members. Trust, a common cause, and often existing personal
relationships provide a preliminary, ad hoc security screen for the core
leadership, but as soon as steps are taken to activate the movement and
reach out to potential donors or new recruits, a much more rigorous
and standardized security policy must be in place. Even in the absence
of having previous experience with insurgencies (as is common among
new insurgent leaders), leaders of a new movement can easily acquire
guidelines and "best practices" from the broad field of academic and
military research on the topic of security.

Common Bonds

Although it seems intuitive that a common cause and common ideology will be the determining factors that draw new members into an
underground organization, research shows that people may join the
same movement for a variety of motives, and this disparity can serve
to undermine the cohesion and security of the organization (see the
second edition of *Human Factors Considerations of Undergrounds in Insurgencies*). As such, a common bond (religious, ethnic, tribal, etc.) often

provides a much more resilient contribution to organizational unity and security.

The Revolutionary United Front (RUF) in Sierra Leone adopted insurgency principles and formal organizational structures that its leadership learned during indoctrination in Libya. Despite this appearance of military structure and discipline, however, the fact remained that the RUF was composed of mercenaries, student radicals, unemployed youth, discharged junior soldiers, and other people whose identity was based on opposition to "the system" as well as self-preservation. In this regard, it was no surprise that internal power struggles emerged almost immediately within the RUF leadership structure, to include two significant events in 1993. The first event involved the execution of RUF's second-in-command and forty other RUF members by other top RUF leaders due to personal rivalries. The second event was the torture and execution of twenty-five members of the RUF's First Battalion (who were mostly ethnic Temne from Northern Sierra Leone) by RUF leaders and rebels who were ethnic Mende from Southern Sierra Leone and who wanted to shift the leadership structure of the RUF to those of southern ancestry. These internal purges as well as actual combat losses eliminated some of the most popular and competent leaders within the movement and decimated the limited military expertise of the RUF during the early years of the conflict.[2]

Similarly, the New People's Army (NPA) in the Philippines conducted numerous purges of its members, including the Second Great Rectification, during which thousands of members were killed on the basis of the accusation of these members being "deep penetration agents." Yet in reality, these purges were an effort by one NPA faction to regain control of the Communist Party of the Philippines NPA from the autonomous regional commission and from larger NPA formations that had recently risen to power. The cohesiveness and security of NPA units depended on comradeship (i.e., trusting your comrade with your life), which suffered greatly because of the purges.[3]

Compartmentalization

Compartmentalization is the division of an organization or activity into functional segments or cells to restrict communication between them and prevent knowledge of the identity or activities of other segments except on a need-to-know basis. This security measure minimizes the danger of compromising the organization and is accomplished by limiting contacts between superiors and subordinates to include lateral liaison with leaders in other cells (see *Chapter 1: Leadership and Organization*). Thus, if captured, a member cannot expose anyone other than his fellow cell members because the use of code names protects the identity of others of whom he may have heard or met. However, compartmentalization is difficult to maintain because of the early practice of recruiting friends or relatives, members who know other members, and leaders who may also know more than they should about the organization. While it is not possible to completely eliminate this situation, issuing orders warning against inquisitiveness by average members or indiscretions on their part will help to lessen the danger and reduce the proliferation of information.

This was the approach taken by the Sendero Luminoso, a socialist insurgency in Peru, which implemented a cellular tactical structure, with each cell generally composed of ten or fewer personnel. Only the leader of that cell knew other members outside of that cell, and then, only by aliases. Identities of members were carefully protected, and all official communication between cells was conducted by the cell's commander. This process of limiting contact between cells was known as "regulated liaison." However, when the Peruvian security forces began to target the cellular structure of the Sendero Luminoso organization, they were able to destroy the effectiveness of an entire cell—and often more—simply by focusing on the leadership, not the individual members. Because none of the other members of a cell had any contact with other cells, their effectiveness was generally gone when their leader was gone.

Use of Couriers

With the advent of cellular communications and the heavy use of e-mail for daily communications, underground movements began to fall victim to the advanced electronic surveillance capabilities of security forces that targeted these electronic communications. In response, these undergrounds reverted to their older, traditional reliance on couriers and became more disciplined with their use of brevity codes and aliases.

Aware of U.S. communication interception capabilities, the Taliban began using messengers as their primary means of communication; this decision reflected the traditional Afghan practice of making sure a planned operation remains secure. In addition to traditional messengers, the Taliban also communicated using written communication, pamphlets, underground newspapers, and communal messages as well as "night letters." These "night letters" were not only used between Taliban members but also between the Taliban and non-Taliban Afghans. In the case of Taliban to non-Taliban usage, the "night letters" often contained death threats to anyone who dealt with foreigners and served as a way to maintain control of the public. Given the Taliban's goal to revert Afghanistan back to the times of the Prophet Mohammad, the Taliban favored this slow form of communication over the Internet and mobile phones, which were also more susceptible to interception. In fact, because the Taliban were weary of civilian informants, they even banned cell phone usage because cell phones were a means for private citizens to call in tips on the Taliban's plans. As a result, in March 2008, the Taliban ordered that cell phone companies in Kandahar "suspend service from five in the evening to seven in the morning so that the Taliban could operate safely during those hours."[4]

The use of a courier, however, does not always guarantee the security of insurgency members. In the case of the largest manhunt in U.S. military history, the personal courier of Osama bin Laden became the only real breadcrumb in the lost trail that eventually led to bin Laden's hideout in Abbottabad, Pakistan. Like the Taliban and other insurgent movements, the leadership of Al Qaeda was acutely aware of U.S. electronic surveillance capabilities and experienced numerous close calls as well as successful strikes by American forces using cell phones to locate targets. Bin Laden therefore severed his direct electronic ties to the outside world and relied on a very tight, trusted inner circle to facilitate his communications, to include a dedicated courier. The identity of this courier (Abu Ahmed al-Kuwaiti) was confirmed by U.S. Central Intelligence Agency analysts in August 2010 after years of trying to determine this link between bin Laden and the outside world. The analysts determined that al-Kuwaiti drove a white sport utility vehicle that had a graphic of a white rhino emblazoned on its spare tire cover and began tracking the vehicle with satellite imagery. It was soon determined that this vehicle and its driver spent a large amount of time at a large concrete compound in Abbottabad.[5]

SCREENING OF NEW MEMBERS

Undergrounds exercise caution when inducting new recruits in order to protect against infiltration by counterinsurgency penetration agents. Recruits, therefore, are not fully accepted until a check can be made of their backgrounds, including political activities, jobs, and close associates. Moreover, once accepted, most undergrounds require recruits to undergo a probationary period as a further precaution. Prospective recruits identified as enemy agents are often eliminated; others whose backgrounds do not clearly indicate their reliability are shunned. In cases in which an individual is needed before a background investigation can be concluded, the recruit will be prohibited from coming into close personal contact with cell members until they have been fully cleared. This is usually done by restricting their contact to one member and at places other than the cell's usual meeting places.[6] When the British increased their use of informants and undercover agents within the Provisional Irish Republican Army (PIRA), the PIRA mounted counter-operations against the intelligence network and reconnaissance force by raiding a massage parlor, ice cream shop, and laundromat that were run by the British operation and effectively shut down that part of the British network.

LOYALTY CHECKS AND OATH

Undergrounds often administer loyalty oaths to new members in order to emphasize the seriousness of the movement and the requirement for them to be completely devoted to the movement. If religion is a critical aspect of daily life in that country, the loyalty oath may also integrate key religious references in order to elevate the authority of the loyalty oath and increase the sense of obligation and commitment that the applicant has toward the movement. It is not unusual for recruits to be warned that any betrayal of the movement may be punishable by death.[7]

For the Sendero Luminoso in Peru, all recruitment was initiated by current members, and volunteers would be viewed with suspicion. Accordingly, two current Sendero Luminoso (or Shining Path) members had to vouch for the new recruit. During the initial year of a two- to three-year training process, recruits engaged in simple noncombat tasks, such as distributing propaganda. They also received classroom instruction on Marxist texts and guerrilla warfare as well as indoctrination in Sendero's ideology. After one or two years, recruits began their military training, which included participation in acts of sabotage, such as destroying high tension towers, bridges, and other infrastructure. Recruits also engaged in physical conditioning and training in small arms use, explosives, combat triage, and other specialized guerilla warfare proficiencies. At the conclusion of the training period and after a thorough investigation of the candidate's background, particularly personal associations, a formal determination was made as to whether the recruit would be initiated. If the recruit was accepted, they took an oath before four regional leaders who were hooded to protect their identity. Even after two or three years of training, a recruit would still only possess very limited knowledge of the organizational structure, having only had contact during that time with a handful of other members because the training cadre used a cell-like structure.

Communications Security

As mentioned above, communications security is a fundamental security policy for underground networks. Code words are used to designate places, movements, and operational plans. Code words are also used as a means of recognition between members meeting for the first time or to indicate danger. Use of unsecure communications, such as e-mail and cell phones, is limited. And as much as possible, couriers are used to transfer information, funds, and other assets between cells.

Modern communications and the then-emerging interconnectivity provided by the Internet played a fundamental role in almost every aspect of the formation of the Ushtia Çlirimtare e Kosovës (Kosovo Liberation Army, or KLA). With cellular telecommunications and increased speed of data transmission, it was possible to control certain aspects of the organization from external and noncoerced areas. The use of unsecured cell phones and personal mobile radios was the primary form of communication for intra-unit purposes as well as for external sources reaching into Kosovo proper. During the onset of the major stages in the conflict in 1998, the primary methods for military coordination were the use of unsecure commercial cell phones, faxes, and face-to-face encounters. Every KLA leader assumed that his or her cell phone was being tapped by Serb intelligence services, so all discussions and exchanges were kept to an absolute minimum and strict adherence to operational security was entailed. Places, names, and specific observables were not mentioned because it was expected that operational plans could be easily interpreted by the Serbs.

Even during Sierra Leon's civil war, which was known more for its basic brutality than for any advanced usage of communications, let alone communications security, the RUF received training and operational support from some South African mercenaries, and this assistance led to the use of the old British Slidex encryption code for important internal communications.[8]

Record Keeping

Although record keeping is necessary for any organization, whether clandestine or not, it is important that written records be kept to an absolute minimum in an underground. Only information that cannot be memorized or information that is needed for future reference should be put down in writing to avoid exploitation by the counterinsurgency forces. A worst-case example of what can happen to movements that choose to maintain extensive written records is the case of a World War II French resistance network called the "Cartel," which was sponsored by the British Special Operations Executive (SOE). Andre Girard, leader of Cartel, and some of his colleagues drew up lists of the Cartel's membership; this list contained paragraphs of personal description—names, addresses, appearances, telephone numbers, experiences, specialties, capabilities, discretions, and so on. All this information was written in the clear (i.e., not encoded) and was kept in Girard's office or study. In November 1942, more than 200 of the most important cards were being taken by a courier from Marseilles to Paris by train. During the trip, the courier fell asleep. When he awoke, the briefcase with all the cards was gone—it had been taken by a German intelligence agent. The downfall of the Cartel was therefore assured.[9] A more recent example includes the ability of U.S. counter-terrorism officials to gain access to laptops and hard drives belonging to Khalid Shaikh Mohammed and Osama bin Laden, which have also provided a treasure trove of information on Al Qaeda and its operations.

Underground Discipline

Obedience to security rules and procedures is a critical requirement for the survival of an underground. Members are often required to report all violations and may be subject to disciplinary action if they fail to do so. In this regard, the ethics and "professionalism" of the membership can play a large role in the organization's discipline. Those underground movements that gravitate toward criminal activities to fund the movement (or, in some cases, those movements in which criminal activity becomes a target of opportunity for certain members) may see the dedication and discipline of members deteriorate as they become involved in narcotics, arms, diamond, and other illicit trafficking. Other movements may initially start off with high standards for membership but then slowly relax those standards (and lose discipline) as the movement grows and the need for new members grows while the available pool of quality recruits dwindles. To promote discipline and underscore the importance of security to the movement, some undergrounds execute members for serious or frequent breaches, and traitors are also usually executed to discourage others from betrayal.[10] Finally, some of the more ruthless undergrounds will impose heavy tolls on the local population that supports them if it is perceived that there are people in the population who are collaborating with security forces. Examples are plentiful and include the December 2006 execution of twenty-six men in a Taliban-dominated village west of Kandahar City because these men were accused by the Taliban of cooperating with International Security Assistance Force troops. As an added measure, the Taliban publicly displayed the headless bodies to warn locals that they would face a similar fate if they collaborated with the coalition or government forces.

PERSONAL SECURITY MEASURES

At the individual level, a basic requirement—and skill—of an underground member is to blend into his or her surroundings. The member must establish a normal routine that is consistent with the routine of his or her neighbors in order to avoid attracting attention. A former leader in an anti-Nazi underground movement in Germany stated that "you can't hide from the scientific surveillance of a modern police state, but you can mislead the police. And the best way to mislead them is to live as conventionally and as openly as possible. The more you resemble a normal everyday citizen in every respect, the less apt you are to be suspected."[11] This same philosophy served as a basis for Dr. Ayman Al-Zawahiri and the Egyptian jihadist movement. His obsessive-compulsive attitude (he was detail oriented, rigidly enforced challenges and passwords, consciously avoided of pattern setting, etc.) and paranoia (he assumed the group was under surveillance, exhibited caution in recruitment, attempted to avoid large congregations of jihadists, etc.) were traits that served him and his colleagues well in avoiding the Egyptian law-enforcement and counterintelligence apparatus.

An important aspect of personal security is the establishment of random behavior patterns. Normal, even innocent, behavior patterns may draw the attention of security forces if the activities are repeated in a noticeable pattern. This repetition may also enable the security forces to anticipate future actions and plans. To prevent this possibility, undergrounds will alter their courier routes, living arrangements, meeting places, mail drops, and day-to-day activities.

SAFE HAVENS

The robust intelligence-gathering capabilities of modern counter-insurgency forces have elevated the importance of "safe havens" for the security of underground movements. Recent trends have demonstrated a preference toward ungoverned territories where there is limited government oversight and where bribery and corruption carry the day. A striking example of the benefit an underground can achieve with access to a safe haven is shown with the Fuerzas Armadas Revolucionarias de Colombia (Revolutionary Armed Forces of Colombia, or FARC) in Colombia. President Andres Pastrana was elected in 1998 and assumed that pacification, not military confrontation, would drive the FARC to negotiation. Pastrana even ceded an area of the country as a demilitarized zone, allowing de facto control by the FARC, called the Despeje—an area that was the size of Switzerland. The FARC took this opportunity of localized self-rule to increase its legitimacy in the eyes of Colombians and the international community. Ambassadors from various countries were invited to the Despeje to meet with FARC leadership, further supporting the FARC's position as a legitimate governing body. They also used this expansive safe haven to recruit, train, improve capabilities, and establish governing bodies in local communities. Meanwhile, production, processing, and trafficking of coca and cocaine flourished in the region. The FARC was also reported to use territory in Ecuador, along Colombia's southern border, as a safe haven, for training, and to support illicit business activities associated with coca production. Finally, Venezuela was alleged to have provided safe haven to insurgent FARC fighters and leadership, supporting training sites as well as harboring command and control locations.

In the Philippines, the NPA established safe havens over time in hundreds of areas where the local populace could be depended on for logistics support from food, to medical supplies, to batteries. Sendero Luminoso used the University of San Cristobal de Huamanga and the Ayachucho region for several years as a safe haven, building on the disenfranchised local Indian communities while also recruiting intellectuals from the university to grow a political and military base.

ACTIONS IN THE EVENT OF COMPROMISE

Despite its best efforts to promote security for the organization, the chance always remains that an individual, a cell, or key leaders of the movement may at some point be compromised by counterinsurgency forces. The ability to rapidly identify and react to this compromise is critical to ensuring the overall security of the movement.

An initial indicator of possible compromise can occur when a member has failed to keep an appointment, is not reachable by the cell leader, or has disappeared altogether. Every effort is made to determine whether the member has been captured. If capture is verified—or, as in most cases, if contact is not established and capture must be assumed—all members with whom the member was in contact or about whom the member possessed information must be immediately alerted because it should be assumed that the captured colleague has been forced to talk. Affected members immediately abandon their present identities and living arrangements and move to safe houses until a new identity and dwelling can be arranged. For this purpose, all members should have one or more safe houses known only to them where they can stay, with the types of safe houses varying from the homes of friends, remote rural cottages, or even warehouses.[12]

When a member is captured, the underground must find out what led to his compromise. If it is determined that he was betrayed by someone within the movement, the informer should be quickly identified and executed. The underground must also react quickly to the capture of members who possess damaging information that could seriously endanger the movement. Every effort must first be made to free the member using methods such as bribery, threats, or blackmail. If these methods fail, an armed rescue of the captured member may be attempted. When these options are not possible or unsuccessful, the movement may then attempt to kill the member in order to silence him.[13]

ACTIONS IN THE CASE OF CAPTURE

As part of the indoctrination process for new members, undergrounds will often expose the recruits to a code of conduct for how they should behave if captured. An example of these "rules" is found in a 1935 book for German Communists living in Prague. Among the tenets listed for personal actions when arrested are the following (as cited in *The Underground*, ST 31-202):

- As a matter of principle, I do not divulge names, cover names, personal descriptions, addresses, or places through which comrades could be contacted.

- I never admit my guilt in any offense I am accused of, even if all evidence is against me.

- When I am told that others have already confessed, I do not believe it, and if others have really confessed, I will call them liars, always denying everything.

- If I am tortured or beaten, I will let them kill me, torture me to death, rather than betray my organization, my comrades.[14]

An additional part of the training for new members is instruction on what to expect if captured in order to better prepare members for any interrogation techniques they may encounter in prison. Common interrogation and interview techniques include the insertion of a deception agent into the same cell to attempt to gain the confidence of the member and obtain information from him. Another method involves demoralizing a prisoner by using information already known about him or the organization in order to make him believe that he was betrayed, thereby undermining his loyalty. Security forces may also attempt to gain information by exploiting a member's ego by praising his exploits for their daring and asking him how he accomplished those exploits. Another common technique is to promise leniency or amnesty to the member in exchange for information.[15] If these techniques fail to achieve the results desired by the security force, torture may be used; therefore, most organizations to assume that torture will eventually be used, that the captured member will eventually be broken of his will, and that information will be divulged. For this reason, underground training for new members will emphasize that members may likely be broken but that they must strive to resist divulging information for approximately forty-eight hours in order to allow sufficient time for impacted members of the movement to be notified and proper security measures taken.

ENDNOTES

[1] *The Underground*, ST 31-202 (Fort Bragg, NC: U.S. Army Institute for Military Assistance, 1978).

[2] Jerome Conley, "The Revolutionary United Front (RUF), Sierra Leone," in *Casebook on Insurgency and Revolutionary Warfare, Volume II: 1962–2009*, ed. Chuck Crossett (Laurel, MD: The Johns Hopkins University Applied Physics Laboratory, 2009), 532–558.

[3] Ron Buikema and Matt Burger, "New People's Army (NPA)," in *Casebook on Insurgency and Revolutionary Warfare, Volume II: 1962–2009*, ed. Chuck Crossett (Laurel, MD: The Johns Hopkins University Applied Physics Laboratory, 2009).

[4] Sanaz Mirzaei, "Taliban: 1994–2009," in *Casebook on Insurgency and Revolutionary Warfare, Volume II: 1962–2009*, ed. Chuck Crossett (Laurel, MD: The Johns Hopkins University Applied Physics Laboratory, 2009).

[5] Nicholas Schmidle, "Getting Bin Laden," *The New Yorker*, August 8, 2011, http://www.newyorker.com/reporting/2011/08/08/110808fa_fact_schmidle.

[6] *The Underground*, 56.

[7] Ibid.

[8] Al J. Venter, *War Dog: Fighting Other People's Wars* (Drexel Hill, PA: Casemate, 2003), 53.

[9] *The Underground*, 67.

[10] Ibid., 68–69.

[11] Ibid., 69.

[12] Ibid., 89–90.

[13] Ibid., 90.

[14] *The Underground*, 90.

[15] Venter, *War Dog*, 90.

CHAPTER 9.

SHADOW GOVERNMENT

CHAPTER CONTENTS

Summer Newton and Robert Leonhard

INTRODUCTION

The tactical use of governance activities to influence civilian behavior or fulfill operational objectives by undergrounds is captured by the term "shadow governments." As the term implies, shadow governments are formal or informal governance activities oftentimes operating in tandem with those of the incumbent state government. Because of the predominance of the state as the accepted legitimate form of political organization, shadow governments mimic the attributes and functions of the nation-state and, in effect, represent a "counter-state."

Insurgent groups implement shadow governments in pursuit of a number of objectives. Oftentimes, shadow governments are a reflection of the important objective to legitimate the authority of the insurgent group and gain popular support. However, insurgent groups may also use shadow governance activities to undermine the official government or to extract crucial resources. Shadow governments differ from one another in a number of ways, including their institutional complexity, effectiveness, and objectives. A number of factors are thought to account for these variations, including the political context as well as a group's internal dynamics and objectives.

Lastly, changes in the nature and termination of conflict in the post-Cold War era have made the public component, or political arm, of insurgencies increasingly important. Oftentimes, the public component of the insurgency is involved in shadow governance activities. As negotiated settlements have become more frequent, so too have opportunities for insurgents to "change their stripes" and enter the political process. Negotiated settlements, however, present challenges to the duration of peace in the post-conflict environment.

THE FORM AND FUNCTION OF SHADOW GOVERNMENTS

Numerous changes in the functions of shadow governments have emerged since the publication of the previous *Undergrounds* edition.[1] The older work only very briefly discusses shadow governments as a function of insurgent undergrounds to secure popular support, distinguishing between internal and external, or exiled, shadow governments. Popular support remains an important component of insurgent and counterinsurgent strategy. However, since the publication of the previous edition, the nature of conflict has changed considerably.

133

Today, combatants are more likely to engage in civil war than interstate war.[a] As a result, an insurgency is more likely to establish an internal shadow government, or one within the boundaries of the contested state, in an effort to undermine or wrest legitimacy from the incumbent government than to establish an external, or exiled, government. Additionally, the previous edition discusses shadow governments as a tactic employed by undergrounds to aid in capturing political control once the conflict has ended. However, conflict in the post-Cold War era is increasingly likely to end not in the capture of existing state institutions and center of power, but in a negotiated settlement that demobilizes and transitions the armed group into the political process. These changes result in an expanded role for the public components, or associated political parties, of insurgent groups.

The formation of shadow governments is influenced by the international system in which they develop. Since the dissolution of the Ottoman and Austro-Hungarian Empires after World War I, the nation-state (or more commonly, "state") has been regarded as the primary legitimate form of political organization. The state, as a form of political organization, exerts a powerful normative[b] force that influences how insurgent groups govern and establish political institutions. The important issue of legitimacy is discussed in greater detail below. For now, it is enough to note that many attributes of shadow governments mimic those of the state. Often, insurgent leaders purposively mimic these functions to bolster a shadow government's legitimacy in the eyes of both its domestic and international audiences.[4]

Insurgents are in an "interactive role" with the state they oppose, not only in terms of kinetic activity, but also as an idea, because the state is regarded as the "predominant form of political community" by domestic and international audiences.[5] Scholars have identified a number of attributes commonly used by shadow governments in this competition with the state. Insurgent groups might adopt one attribute or replicate the whole range of attributes. Somaliland, in the northwest region of Somalia, operates as a de facto state, lacking only international recognition to distinguish it from other bona fide state organizations. The ability and desire to adopt one or more state attributes may change during

---

[a] Since 2003, the Uppsala Data Conflict Program (UDCP) has recorded only one interstate conflict, between Eritrea and Djibouti. Of the thirty-six active conflicts recorded in 2009 by the UDCP, all were fought within states. This was also true for the period 2004–2007. However, seven of the thirty-six conflicts were internationalized, or involved troops from external states aiding combatants. The UDCP defines armed conflict as "a contested incompatibility that concerns government and/or territory where the use of armed force *between two parties* results in at least 25 battle-related deaths in a year [sic]."[2,3]

[b] The term "normative" is frequently used in philosophy and the social sciences to describe conditions that are the ideal or the standard by which others are judged.

the course of a conflict. The National Resistance Army, or NRA, a resistant movement in Uganda, adopted governance strategies supporting the civilian population until military pressure stopped them.[6]

Insurgent undergrounds often compete with the government in projecting force throughout a country or region while denying or disrupting the government's ability to do the same. An insurgent group may severely limit or altogether halt the enemy's ability to operate kinetically in the territory as well as control more common criminal elements. An insurgent group's successful extension of force can prevent the incumbent government from ruling its territory, as was the case with National Union for the Total Independence of Angola, or UNITA's challenge to the Angolan authorities:

> Lodged like a bone in the throat, [UNITA] offered a permanent challenge to Luanda's [ruling MPLA[c] party] authority, to its ability to implant policies that might ordinarily have improved the lives of Angola's people. It denied the very title that MPLA had won for itself as the Government of the People's Republic of Angola. Savimbi's campaign . . . meant that the MPLA did not, could not, govern the country.[7]

Likewise, in 1998, when the Fuerzas Armadas Revolucionarias de Colombia (Revolutionary Armed Forces of Colombia, or FARC) took control of five municipalities granted to it by the Colombian government, the incumbent state was admitting its inability to effectively project its power into territory under its jurisdiction. In the period after the turnover, FARC not only blocked government forces from exercising power, but the group was also able to affect a significant drop in criminal behavior such as murders, robberies, and rape.[8] In strongholds where it enjoyed significant support, in addition to making it difficult for security forces to operate (particularly the police), the Provisional Irish Republican Army, or PIRA, took effective responsibility for reducing incidences of criminal activity.[9] Offering security and police administration is the crucial first step in establishing a shadow government and cultivating popular support. This first requires, of course, capturing and holding territory through the use of violence. Perceived strength and beneficence are often both crucial components of "legitimate authority."[10]

Insurgent groups often must battle not only the incumbent government for armed supremacy but also rival insurgent groups. Since the end of the Cold War, intrastate war has increasingly involved more

---

[c] People's Movement for the Liberation of Angola–Labor Party or the Movimento Popular de Libertação de Angola–Partido do Trabalho.

than one viable rebel group challenging the incumbent government. In 2002 and 2003, 30 percent of conflicts involved more than one challenge to the state,[11] and in 2009, 20 percent of conflicts involved more than one challenger.[12] Insurgent groups have an incentive to "dominate, ally with, or destroy weaker rivals" in order to "establish national control by one's own forces."[13] The Liberation Tigers of Tamil Eelam, or LTTE, engaged in a bitter rivalry with other Tamil insurgent groups in the mid-1980s,[14] and the PIRA's domination in the Northern Ireland conflict required its armed defeat of its rival rebel group, the Official IRA, in the early 1970s.[15]

Governance activities are dynamic. Scarce resources and changing political and military contexts can lead insurgents to revaluate governance strategies. Loss of the extension of force in territory, for example, can make it difficult for insurgents to continue with governance activities. The NRA, operating in Uganda, offered a series of services, including health care and security, to the civilian population in its liberated areas. As its hold on those areas deteriorated during the war, the NRA evacuated civilians to safe pockets in the Luwero Triangle, still encouraging civilians to maintain the democratic village councils it established in its safe areas. Eventually, as its position became more precarious, the NRA was forced to terminate all ties with the civilian population in order to free the group from allocating resources to civilian defense. The NRA demanded that civilians leave the war zone. NRA only resumed governance activities when its military position vis-à-vis the incumbent government improved considerably.[16]

Shadow governments often adopt other attributes of states, including national identity and legitimacy, revenue generation, and the provision of social services.[d] National identity and legitimacy address the extent to which residents regard a shadow government as the legitimate ruling power over and above the incumbent state or other competing authorities. Many states within the international system, particularly those in the developing world, lack this basic state attribute. Oftentimes colonial and imperial powers created states without regard to the ethnic diversity or historical borders its residents found meaningful. Some insurgent groups, such as the Eritrean People's Liberation Front, or EPLF, use this technique to their advantage; the EPLF used the Ethiopian threat to emphasize an Eritrean identity over and above significant extant religious and ethnic cleavages. Today, Eritreans evince a strong sense of nationality often missing from many sub-Saharan African states.[18]

---

[d] These attributes, along with the extension of force, are discussed by Spears,[17] although Spears uses the label "Infrastructure and Administration" as opposed to our "provision of social services."

Fiscal policies are often thought to be in the domain of government, but insurgents also need to generate revenue, which they often do through taxation. The creation of a legitimate state is "intimately bound with the creation of fiscal institutions that are acceptable to the majority. . . ."[19] Occasionally, insurgents may be more effective at gathering taxes than the government itself. Local populations may prefer the tax efforts of insurgents. Sympathetic populations may prefer to pay rather than evade taxes, as was the case during the Kosovo insurrection when the Kosovo diaspora in Germany contributed funds through a well-organized, if informal, payroll tax. Even populations expressing little support for insurgents may prefer insurgent to government taxes, particularly if the group is perceived as less predatory and offers more security than the government.[20] Shadow governments and insurgent groups gather funds in a variety of ways, whether though taxation, voluntary contributions, extortion, kidnapping, or other criminal activity, like drug trafficking.[c] During the Cold War, insurgencies frequently received funds from outside states. Today, insurgencies need to be more creative in generating income, leading an increase in the importance of control of "revenue-generating regions" and the potential for greater victimization of civilian populations. UNITA faced challenges after its primary source of income, foreign assistance, evaporated after the end of the Cold War. Afterward, the group's revenue-generating strategies transformed, relying heavily on territorial control of diamond-rich areas as well as taking advantage of other commercial activities in the group's territory—these strategies sometimes generated as much as $5 million a month. Despite an agreement to move toward centralized government, when government forces encroached on the diamond-rich territories, UNITA returned to violence.[21] Extraction of resources from a target population is, for some insurgent groups, the predominant objective of its governance activities.

Lastly, shadow governments provide needed social services— "charitable acts, public services, and infrastructure development"[22]—to the civilian population. Shadow governments might build roads, telecommunication networks,[23] or other municipal services as well as offer education and health services. Hizbollah has been especially effective in this regard. The group's Social Service Section used half of Hizbollah's 2007 budget for social services, which were delivered to the group's mostly Shia constituents.[24] The Section is divided into six subgroups supporting various needs of the community, from reconstruction, to providing for the families of martyrs, to women's welfare, to education. Hizbollah's social service efforts, such as the reconstruction of homes and structures damaged by the 2006 war with Israel, far outstrip those

---

[c] See Chapter 4 for a discussion of revenue generation.

of the Lebanese state, which has done little to improve infrastructure in Shia neighborhoods since the 1900s.[25] Hizbollah's efforts, largely financed by Iran, have reaped handsome rewards in political and military clout in the Levant.[26]

GOVERNMENT VERSUS GOVERNANCE

For the purpose of this chapter, it is most helpful to conceptualize shadow governments not just as a series of formal political institutions erected by insurgents, but as a tactic employed by numerous non-state actors, especially insurgent movements, to control or influence target populations in pursuit of their objectives. Governance is not synonymous with government. While both terms refer to "purposive behavior," "goal-oriented activities," and "systems of rule," it is only government that is backed by a formal authority and police powers to ensure implementation and compliance with civil laws.[27] Governance fulfills a similar role as government, but it can do so without formal laws or ways of enforcing compliance and without institutions created specifically for carrying out the duties of the state. In other words, taxation, education, and security may still occur but without recognizable formal institutions, such as an Internal Revenue Service, a Department of Education, or formal police forces, to do the job.[f] Insurgents may carry out governance activities in a more ad hoc or informal manner. It is possible, as in the case of weak states, to have formal institutions of government without the benefit of governance.[29] That is, it is possible to have formal institutions of government that are either unwilling or not able to carry out the duties for which they were created.

SHADOW GOVERNMENTS IN THE
INTERNATIONAL SYSTEM

It would be difficult to understand why and how shadow governments form without also understanding the political, social, and economic stresses many states are undergoing across the globe. Shadow governments are only one example of political organizations forming, and gaining legitimacy, outside the traditional state. Shadow governments are a particular problem for weak states. Many of these weak states in the developing world are said to have "artificially constructed" boundaries, or borders created by European colonial powers without

---

[f] Nelson Kasfir describes governance as "the range of possibilities for organization, authority and responsiveness created from the daily interactions between guerrillas and civilians."[28]

regard to traditional or existing economic, geographic, or ethnic group-
ings. However, undergrounds have also formed shadow governments
in states with deeper historical roots, like Lebanon and Colombia.[30]
Combined with ineffective, and often corrupt, government institutions
and lackluster economic performance, these states struggle to obtain
legitimacy and provide the security, prosperity, and political identity
citizens expect.

The tendency to associate order with the nation-state and centralized rule has led scholars and policy makers to see anarchy in territory
lacking effective central governments. Often called "failed states," "weak
states," "ungoverned territories," or "black spots," the absence of central rule is sometimes mistaken for a complete lack of governance. The
absence of centralized rule, however, does not necessarily indicate that
there is no governance. Local governance structures often flourish in
these areas. Somalia, the archetypal failed state, has a number of local,
regional, and municipal governance institutions that operate without
any interference or assistance from the impotent central government.[31]
It is probably more appropriate to discuss some of these regions not as
"ungoverned spaces" but as "zones of competing governance" where governance is conducted primarily at the local level, often in competition
with the centralized state for sole, legitimate authority over territory.[32]

More broadly, the proliferation of shadow governments and other
political organizations outside the state is part of a larger trend of shifting patterns of state authority. The political authority of the state is
under serious challenge, leaving many weak states scattered across the
world. Weak states are those that have difficulty effectively controlling
and implementing policies in the territory technically within their
international borders.[g] In Somalia, for instance, even before the fall of
President Siad Barre's regime in 1992, his opponents derisively referred
to him as the "mayor of Mogadishu," pointing to the lack of control his
government had over the rest of the Somali state.[33]

This weakened capacity of states, and the seemingly inability of
many, even in the developed world, to control forces profoundly affecting their future, is attributed to a host of factors, but many converge on
the effects of globalization. Globalization is difficult to precisely define
but generally refers to the increasing interconnectedness of nations
and peoples across the world.[h] Interconnectedness appears in political, economic, military/security, and cultural spheres but still within

---

[g] Scholars differ in both how they define weak states and how they measure them.
State weakness is also considered a contributing factor to the outbreak of civil wars. Many
scholars research the attributes of state weakness that contribute to political violence.

[h] There is a "broad consensus" that the primary driver of globalization is the market
economy.

an international system based on the nation-state. Globalization is usually depicted as a process fueled by the spread of neoliberal economic principles[i] and technology.[j] The increased ties and access to information have created problems for states with regard to their distribution of goods and power among their citizens. Access to satellite television, for instance, increases expectations of wealth and quality of life more quickly than states are able meet them. Meanwhile, the highly integrated global economy encourages many regimes to implement policies[k] that might strengthen their position (or merely keep themselves afloat) in the global marketplace but also might cause them to cede control of their economic future to the whims of the global market or supra-national economic organizations like the International Monetary Fund (IMF).[37] Researchers studying conflict now know that unmet expectations alone do not inevitably lead to the outbreak of war, but, like an impaired immune system, they can make a state more vulnerable to politically motivated violence.[38] States face challenges from a variety of actors—shadow governments, international organizations, and criminal organizations, among others—and such challenges limit the state's exercise of sovereignty.[39]

The prevalence of weak states problematizes state sovereignty. International recognition of statehood, often called "juridical sovereignty," does not necessarily mean that the state possesses "empirical sovereignty," or the attributes normally associated with a state, such as the extension of force or provision of social services. Some shadow governments, such as Hizbollah or Somaliland, evidence more empirical sovereignty than the state government. Yet, political entities like Somaliland often seek juridical sovereignty, or official statehood, in vain. In other words, empirical sovereignty is neither a necessary nor sufficient condition for international recognition of statehood. Unlike the shadow governments described in the previous *Undergrounds* edition, which viewed shadow governments as a means to gain centralized political control of a state after the cessation of conflict, shadow governments established

---

[i] Neoliberal economics, or the "Washington Consensus," is characterized by free markets with limited regulation, protection, and intervention by state or international authorities.[34]

[j] As Steven Metz observes, "Artificial and increasingly fragile states are pummeled by globalization, interconnectedness, and the profusion of information."[35]

[k] Such policies, designed to increase economic competitiveness, include "tight money, small government, low taxes, flexible labor legislation, deregulation, privatization, and openness all around." The policies required for global economic competitiveness, however, also restrict a state's, and citizens', ability to direct its economic policy.[36] Additionally, austerity measures adopted by states like Greece accompanying IMF bailouts generally require governments, for instance, to forgo "big government" developmental policies favored, and oftentimes direly needed, by its citizens. Protests, riots, and general unrest, such as those that occurred in Greece in 2011, can be the result.

by insurgent groups now rarely achieve international recognition, their capabilities and legitimacy among their constituents notwithstanding.[40] The remaining uneasy tension between shadow governments and recognized states points to an international system characterized by "durable disorder." Political authority appears to be shifting away from the centralized, hierarchical state-based model to one that is more dispersed and ambiguous. The result is stretches of territory existing in a sort of limbo between sovereignty and subordination to existing states.[41] Statehood, however, does continue to provide benefits for those political entities able to achieve it. Statehood can open avenues to resources unavailable to unrecognized shadow governments, such as access to international assistance, and for most insurgencies, the achievement of statehood is the crown jewel in their quest for legitimacy.[1]

OBJECTIVES OF SHADOW GOVERNMENTS

Shadow governments vary widely in their institutional makeup, as described above, but also in their objectives. Briefly, shadow governments are typically engaged in the business of "killing, stealing, or serving."[42] Variation exists between insurgent groups and during the life span of an insurgency in response to internal and external environmental changes. Research indicates that in the initial phases of conflicts, insurgents commit few if any resources to developing governance systems, whatever their eventual interactions with civilians. Once territory is captured, some groups opt to eliminate individuals residing therein rather than controlling or influencing their behavior. Insurgents might force the migration of populations or outright kill them.[43] In 1994, Hutu-led paramilitaries, like Interahamwe, systematically eliminated Tutsi populations in communities throughout the country over a several-month period, resulting in some 800,000 casualties.

Contemporary counterinsurgency doctrine, encapsulated in FM 3-24,[44] itself heavily influenced by classical conceptions of insurgency and counterinsurgency, identifies the "hearts and minds" of the civilian population as the crucial battlespace for insurgents and counterinsurgents, even over and above the targeting of armed combatants. Pinpointing civilian populations as the center of gravity assumes gaining popular support is a critical tipping point in armed struggle. Che

---

[1] It should be noted, however, that not all shadow governments actively seek sovereignty. Some, like Somaliland, are secessionist and as such do actively pursue statehood. Others, like Hizbollah, appear to be content with their ambiguous standing with the state, although they may participate in, but not wholly dominate, the political process. Others seek to gain control of the existing state's center of power, rendering a shadow government unnecessary.

Guevara and Mao Zedong, two giants in the development of modern insurgent practice, both emphasized the necessity of carefully cultivating popular support and establishing governance activities and structures to do so. In this regard, the armed struggle for supremacy is a profound struggle for legitimacy, or the consent of those being ruled. Such conflicts are thought to be driven by grievances, or genuine concerns over perceived injustices in political, social, or economic matters. However, it is now commonly argued that some conflicts are motivated more by personal interests, especially the accumulation of personal wealth, than attempts to redress societal grievances. The governance activities of these predatory groups are more often on the "killing and stealing" end of the spectrum than the "serving" end.

WAR AS ECONOMICS BY OTHER MEANS

Some social scientists speculate that "greed" is a significant factor promoting conflict. Countries with abundant, exploitable natural resources, such as drugs, diamonds, or timber, are thought to be at higher risk for conflict.[45] What is not entirely clear is why natural resources have such a profound effect on the likelihood of conflict. The potential revenues might provide the motivation, and means, for insurgents to fight, but natural resource revenue can also enhance the odds that insurgents will be successful, or at least survive, by financing the rebellion. Most recently, the research on the natural resource phenomena points to the importance of where the resources are located. When the natural resources are located within the conflict region, civil wars tend to be of longer duration.[46]

Some civil wars today are characterized as a type of "new war," in which insurgent groups make greater use of indiscriminate violence, resulting in more civilian deaths, and predatory extraction of resources, leading to more civilian victimization. These "new wars" are contrasted with "old wars" as defined by Mao Zedong's dictums regarding the necessity of cultivating popular support. In the new wars, insurgents are more likely to capture territory through population displacements, whether through forced migration, killing, or both, eliminating resistance through violence and intimidation rather than through the careful wooing and mobilization of local populations.[47] These changes are thought, in part, to be due to the economic incentives to use conflict as a source of revenue generation. In other words, this new brand of insurgent is more interested in the generation of loot than justice. Insurgents might finance operations through natural resources, external patronage, or international linkages in gray markets aided by market deregulation and globalization rather than through popular support.[48]

Insurgents are then able to use coercion and violence as their primary tools of control and influence. Groups pursuing conflict in these "new wars" use shadow governments not for gaining popular support but in order to more efficiently extract resources from the land and local population. Many armed combatants in Africa's civil wars have either wholly supplanted or merged with criminal networks.[49]

One of the preeminent examples of this new insurgent, and the variation of insurgent tactics over the lifetime of an organization, is UNITA. UNITA was initially a legitimate opposition group driven by Maoist practices struggling to overthrow the kleptocratic and ethnically exclusive MPLA regime. The group sought greater democratic and ethnic representation, especially for the majority Ovimbundu.[50] Its initial strategies aimed at eroding the authority and legitimacy of the Angolan state. As the MPLA continued to survive, UNITA redirected its efforts to extracting resources, including ivory, timber, gold, and, particularly, diamonds, and wealth accumulation. Its strategy evolved, as one researcher and first-hand witness acerbically noted, into a guerrilla force "whose primary objective is inflicting unrelenting and indiscriminate suffering upon defenceless civilian populations while obliterating all infrastructures as a means to render the country ungovernable."[51] The discovery of diamond mines in areas under UNITA's control is partially responsible. The resulting wealth allowed the group to acquire a considerable cache of weapons and other supplies needed to continue its armed struggle. As it became apparent to the group that the MPLA could not be dislodged by either "ballots or bullets," UNITA instead focused on creating enough disorder in the country to continue its diamond-mining efforts. Through violence and intimidation, UNITA drove away rural populations to government-controlled urban centers to facilitate diamond extraction. The trend is apparent in other African countries such as Sierra Leone and the Democratic Republic of Congo in addition to Angola, where effective governance of civilian populations is increasingly regarded as unduly burdensome when groups like UNITA can instead "enrich themselves without the political and administrative costs of governing."[52]

UNITA's remaining governance activities are oriented toward extraction of resources from its territory, including natural resources like diamonds. Diamonds, rather than cash or bank deposits, are the primary means of wealth acquisition for the group. UNITA acquires diamonds in a number of ways. UNITA representatives extract "taxes" from the production of miners working in diamond mines in areas under the group's control. The tax is paid either in rough diamonds or sometimes cash. Moreover, UNITA offers licenses to diamond buyers to operate in its territory in exchange for commissions.[53] UNITA also accumulates wealth through taxation and extortion of commercial activity in its territory. The group may have earned as much as $5 million a month charging landing fees of about $2,000 to $5,000 for "aircraft bringing in food, medicines, mining equipment and other commercial commodities."[54] UNITA also maintains representatives and offices abroad. Some of the offices operate with the blessing and protection of the heads of state, usually in African states, while others operate without the cooperation of the host country, typically in Europe and North America. The offices are used to "cultivate nongovernmental support within the host country or facilitating important political or commercial activities there."[55] Market deregulation and globalization have made it easier for groups like UNITA to make the "parallel or gray international linkages necessary for survival" and characteristic of these so-called war economies.[56]

LEGITIMATING AUTHORITY

Insurgent groups might also use governance activities to alter the relationship between the incumbent state and its population. As one researcher notes, "rebel governance requires a complicated relationship with a defined civilian population that implicates bigger questions about the nature of authority, legitimacy, and power itself."[57] Legitimacy, or the consent of a population that a political organization has the right to expect and enforce its obedience, is at the core of this authority relationship.[m] The failure of the state to uphold its obligations, whether security, social services, or protection of basic human rights, erodes legitimacy and opens avenues for alternative forms of political order, like shadow governments, to reconstruct the authority relationships to their benefit. Although the issue of legitimacy is widely thought to be an important factor in effective government, it is difficult to scientifically "prove" its role. In studies of guerrilla governance[n] in Latin America, one researcher found that guerrillas were indeed most effective in regions where the legitimacy of the government had declined because of breaches of the social contract, whether in terms of security, justice, or social services. If, for example, residents were still

---

[m] This conception of legitimacy is based on an understanding of the state as a political organization formed through a social contract, a central tenet of Western liberalism. In the social contract, legitimate political authority originates in the consent of those being ruled while outlining a reciprocal relationship of mutual obligations and rights. Today, many citizens expect, and often demand, that the state uphold its end of the bargain by providing security, basic goods and services, human rights, and a meaningful political identity, among others. It is important to note, however, that legitimacy has, and still does, arise from other sources, including ancestral or religious authority.

[n] An alternative term Wickham-Crowley uses to describe shadow governments.[58]

recipients of government provisions, the population responded with "indifference, resistance, and even hostility to the appeals of guerrilla proto-governments."[59] Through the provision of governance activities, guerrilla groups were able to be recognized as the legitimate sovereign authority over and above that of the incumbent government.[60] Wickham-Crowley goes so far as to claim a striking "cultural regularity" of this understanding of legitimate authority.

Sovereignty has several important dimensions, including juridical and empirical sovereignty discussed above. Juridical sovereignty bestows significant benefits to political organizations that are appealing to insurgent groups. In this regard, shadow governments are not playing solely to domestic audiences but to international ones as well.[o] Provisioning social services, for example, can be a tactic employed by insurgents as a "stepping stone to the bigger prize of recognition by the international community—an outcome that few ever achieve."[61] Political organizations can also exhibit empirical sovereignty—the capacity to fulfill the functions normally attributed to states, such as those discussed in the first section. However, whereas juridical sovereignty is necessary for statehood, empirical sovereignty is neither necessary nor sufficient. That is, as demonstrated by weak states, states are not required to possess empirical sovereignty to maintain juridical sovereignty, although they can certainly face challenges for the right to rule by domestic, and increasingly less often, external actors. Nor are political organizations with apparent empirical sovereignty automatically considered states. However, like shadow governments, political entities seeking juridical sovereignty, such as Somaliland, often pursue and market their empirical sovereignty to convince international audiences of their worthiness to be officially recognized as a state in the international system.

The last dimension of sovereignty is the most curious and perhaps the least understood. Political entities project power not just through force, provision of goods, or international recognition but also through iconography associated with the nation-state. Anthropologist Clifford Geertz was among the first to identify the state as "theatre" in a study of nineteenth-century Bali in precolonial times. His careful analysis provided evidence that the Balinese state's authority was established not through force but through symbols, rituals, and myths.[62] Symbols associated with the state can act as powerful motivators for consent among populations and demonstrations of sovereignty to international audiences. Shadow governments will often invest resources in mimicking

---

[o] International audiences can also include diaspora communities, particularly when those communities are an important source of financial and human resources for the organization.

the trappings and symbols of the state that otherwise have no apparent value or pertinence to military goals. These practices might include developing costumes and uniforms for military and state personnel, national currency (even if it has no local value), flags, anthems, or ceremonial burial grounds.[63]

GOVERNANCE EFFECTIVENESS

Shadow governments vary not only in their tactical objectives but also in regard to how effective they are in carrying out governance activities. Shadow governments may be effective or only partially effective, or, in some instances, attempts at shadow governance might be an overall failure. The LTTE and Hizbollah are widely recognized for their success in meeting the needs of the local population while boosting the influence and standing of the insurgent group among civilians. Effective governance is the capacity of an insurgent organization to meet three critical requirements in territory under its control: (1) development of a force capable of policing the population, offering sufficient stability to make governance possible; (2) development of a justice or dispute-resolution system, either ad hoc or formal; and (3) the provision of social services in addition to security. The first crucial step undergrounds undertake in establishing shadow governments, after capturing territory, is ensuring the security and stability of the region. Insurgent groups providing security, but few or none of the other critical requirements above, exhibit partially effective governance. Two abbreviated case studies follow—the first describes the effective shadow governance activities of the LTTE, while the second describes the failed attempt at shadow governance by the Rassemblement Congolais pour la Democratie-Goma, or RCD.

One of the most effective shadow governance activities systems is that established by LTTE in Sri Lanka. Both the political and military wing of the LTTE was under the authority of a single commander. The political wing's ministries included finance, justice, protection, economic development, health, and education. LTTE representatives oversaw the implementation of the group's governance directives in each of its territorial districts. The shadow government's relationship with the incumbent Sri Lankan state was particularly unique. Both the incumbent government and the LTTE vied for legitimacy among domestic populations and international audiences, through the provision of services. As a result, the competing governments formed a symbiotic relationship whereby they worked jointly to provide health and education to local populations. In LTTE territories, governance activities were conducted under the auspices of a dual authority—an

LTTE representative and a Sri Lankan representative. The LTTE benefitted from the relationship because it was able to meet the demands of the residents without taxing LTTE resources, and the Sri Lankan government benefitted because it was able to maintain a hold, however tenuous, on the population living under LTTE control.[64]

Like most insurgencies, the LTTE first established an effective policing and justice system that sought to "normalize" life for civilians in its regions. Its police force eventually grew to over 3,000 officers and became a legitimate and respected institution among residents. An expansive judiciary not only mediated disputes among residents but also acted as a source of revenue for the civil administration through land courts that instituted annual property taxes. The taxes generated steady income for the LTTE, particularly from the wealthy diaspora concerned about property they still owned in rebel areas. Moreover, LTTE set up a respectable legal system, which included elements of Sri Lankan penal code and Tamil cultural norms, after more informal and ad hoc measures generated complaints. The system provided the populace with swift justice, and LTTE took pains to ensure it kept corruption to a minimum.[65]

In terms of providing social services, the LTTE was more involved and effective in education than health services. The group faced numerous constraints in establishing health care infrastructure, including an embargo on medical goods and the flight of highly trained professionals upon whom health care depends. International aid organizations offered basic health care, mainly through mobile centers that oftentimes lacked physicians. Residents with serious health conditions typically sought care in government-controlled areas.[66] The LTTE met more success in providing education. Tamil families traditionally place a great deal of importance on the education of their children as a path to social mobility. The Tamil Eelam Education Council was tasked with carrying out education tasks in concert with the government provincial representative. The result was an impressive continuation of the educational system despite interruptions due to the conflict. Before the cease-fire in 2002, 1,994 primary and secondary schools with an enrollment of 648,000 operated in the province.[67]

International aid and nongovernmental organizations (NGOs), which flowed into LTTE territory after the 2002 cease-fire and the 2004 tsunami, altered LTTE's governance system. The LTTE's civil administration expanded its efforts to facilitate aid money, setting standards for work and where and how to establish projects. Moreover, the tsunami and the influx of aid encouraged greater cooperation between the government and the insurgent administration as they developed joint mechanisms to distribute aid and reconstruction efforts. The government and international NGOs viewed the events as an opportunity to coax the LTTE into the mainstream. The goodwill between the combatants dwindled as the government stalled over a final settlement and the intransigence of the LTTE leadership, who would not accept anything short of full independence.[68] The government soundly defeated the LTTE in 2009 through military measures.

A lack of shadow governance activities can reflect a strategic decision to forgo governance of local populations or a failed attempt at governance. The Lord's Resistance Army, or LRA, in Uganda, for instance, opts not to control territory in favor of greater mobility and as a result makes no effort to govern local populations. The RCD, operating in Congo, is one such case.[69] The group faced numerous internal and external challenges that prevented the execution of its governance strategy. Internally, the leadership was riven between those supporting governance strategies to gain popular support while another division wished to devote scarce resources to strengthening its military capacity. Additionally, well before the war had begun, the Congo state had retreated from territory that came under RCD control,[p] abandoning administration to a diverse set of non-state actors—NGOs, the Catholic Church, and civil society groups. The RCD proved unable to persuade these disparate groups to follow along with its governance project. Civil society leaders "[portrayed] the RCD as a tool of Tutsi domination," and Church authorities also evidenced resistance to RCD rule, expressing similar ethnic sympathies.[71]

Although the RCD attempted to integrate itself into the systems providing justice, health, and education to residents, it failed to do so. Non-state actors such as the Catholic-based Caritas and other humanitarian organizations offered more comprehensive and effectual dispute-resolution services to residents. Many residents viewed the security and justice structures operated by the RCD as tools for revenue extraction because justice typically went to those who could pay the higher bribe. Likewise, churches and international aid organizations provided most of the available health services in the Kivus. Although it took control of the health ministry, the RCD outsourced health services to the disparate groups already providing it to residents upon capture of the Kivus. The RCD's role was limited to monitoring and oversight of NGOs operating in the region and directing the type of health campaigns NGOs embarked on and the areas in which they operated, typically limiting NGOs to areas known to be sympathetic to

---

p The RCD initially attempted to co-opt what was left of state institutions in their efforts at governance but eventually realized these institutions were "incapable of being resuscitated."[70]

the RCD while precluding their operation in areas known to house the RCD's armed competitors.[72]

Lastly, the RCD failed to cultivate legitimacy among the civilian population and develop a strong local base. This failure can be attributed in part to the group's inability to present itself as a unifying, multiethnic revolution and its methods of revenue generation. The RCD's dependence on its external patron, Rwanda, and resulting close ties with Tutsi elements led to the perception that the RCD was a monoethnic organization and the handmaiden of its Rwandan patrons. The perception of the "Rwandan taint" spurred opposition, including armed opposition, to RCD's attempts to fully govern the Kivus. Similarly, RCD's reliance on external patronage and extraction of natural resources precluded any pressing need for the group to cultivate popular support. With easy access to weaponry through Rwanda and Uganda, the RCD's strategy was a quick military victory and regional control through coercive means.[73]

GAINING POPULAR SUPPORT

With the exception of more predatory groups, insurgent groups often consciously adopt a shadow governance strategy in order to shore up popular support in their territories. Popular support may be active, as in the case of auxiliaries who perform activities short of combat, such as running safe houses, storing weapons or supplies, or providing intelligence.[74] Shadow governments frequently make it easier for auxiliaries to perform these activities. The remaining mass base, the followers of the insurgent movement, forms the group of more passive supporters.

Insurgent groups vary not only in whether they pursue popular support as a strategy but also in how they go about capturing popular support. It is not clear how many insurgent groups use popular support as a strategy. One scholar speculates that the groups that actively seek popular support are "most certainly" in the minority.[75] Insurgent groups must capture territory before they can initiate governance activities that cultivate popular support, and many groups engaged in armed struggle find this to be a difficult and unfulfilled endeavor. Some groups that ideologically advocate popular support as a strategy may also find that the practical realities of conflict preclude them from following through on such a strategy.[76] Groups that do seek popular support adopt different tactics to achieve their objective. Some groups opt for voluntary support, while others use more coercive measures. Especially for those groups seeking voluntary support, shadow governance activities often play a large role.

149

Insurgent groups that seek popular support and establish shadow governments face numerous challenges. Shadow governance requires scarce resources and incurs risk. Those encouraging voluntary support need to weigh the importance of governance activities against military goals. When faced with increased military pressure or infringements in controlled territory where governance occurs, insurgents can be forced to choose between protecting civilians and military survival.[77] When its survival was uncertain because of increased military pressure, the NRA, despite its previous commitment to popular support and civilian participation in its governing structures, abandoned its territory and halted all shadow governance activities until its position improved.

Shadow governance also engenders expectations among local populations that can make reversing course difficult.[q] The PIRA's political wing, Sinn Féin, took responsibility for administering a crude justice system in the urban enclaves under its control where regular police forces were unable to operate. Sinn Féin's due diligence in providing needed services to its community did pay off in many ways, helping the party to develop a solid voting base in sympathetic areas.[79] At one point, because of the material and human resources required to maintain the rudimentary justice system, the PIRA opted to halt its activities. Shortly thereafter, public pressure led the Provisionals to reverse the decision.[r] Having only tenuous control over sometimes brutal punishment squads and lacking prisons and low-caliber weaponry, the Provisionals administered punishments that often had gruesome results, and this took a toll on the group's popular support. Indeed, Sinn Féin was acutely aware of the repercussions of tackling the crime problem in its communities. In an early internal document, a Sinn Féin representative cautioned about the need to carefully craft its criminal system to match the values and expectations of the community.[81] However, failing to restrain criminal behavior provided inroads for the police force to re-enter Provisional enclaves, yet continuing the governance activities also left a trail of bitter victims and their family and friends willing to act as informants.[82] As the PIRA transitioned to a more political strategy in the 1980s, the negative effects of its brutal justice system became more apparent, making it difficult to attract the support of moderate populations.

---

[q] Kasfir observed similar difficulties in the NRA's management of the governance activities in its territories.[78]

[r] The military and political wing of the PIRA had reasons for disliking the necessary, but troublesome, governance activities associated with curbing criminal behavior. The military wing thought they brought ill repute to the movement and wasted resources. For members of the political wing, in addition to tarnishing the movement, they also made it difficult to gain political support among more moderate populations.[80]

As evidenced in the NRA's operations in Uganda, cultivating popular support through governance offers operational advantages. Yoweri Musveni, leader of the NRA in Uganda, clearly articulated the necessity of civilian collaboration, whether in providing food, intelligence, or shelter, for the execution of NRA agenda. Musveni formed local governments, or resistance councils (RCs), that included noncombatants in areas under NRA control.[83] The RCs originated in the clandestine committees that the NRA selected as trustworthy contacts to supply the group with needed food, recruits, and intelligence.[84] At the onset of the conflict and their arrival into the Luwero Triangle, the NRA had little existing support on which to build and was regarded as an outsider by most residents there.[s] The group's first priority was establishing supply networks and mobilizing the community,[t] making contact first with influential villagers, many of whom went on to become heads of clandestine committees and elected members of the RCs.[87] Attracting support required tailoring the NRA's professed nonsectarian ideology to more ethnically based rhetoric important to most in the region.[88] NRA leaders also carefully prescribed how combatants were to interact with the local population. Stealing was not allowed, and NRA leadership opted to leave food collection entirely in the hands of the village committees—policies that encouraged more voluntary participation and support.[89] The NRA's changes ensured that locally elected members governed their communities. In exchange, NRA offered civilians security by developing early warning systems alerting villagers to approaching enemy soldiers and providing health care services to prevent and treat infectious diseases and other aliments in the liberated zones.[90] As discussed elsewhere, when faced with overwhelming military pressure, the NRA interrupted its governance activities to ensure the survival of the group, but it reinitiated the RCs once its military position improved.

EXPLAINING VARIATION IN SHADOW GOVERNMENTS

This chapter has discussed several variations in shadow governments, including variations in institutional complexity, objectives, and

---

[s] This contrasts with indigenous insurgent groups such as the PIRA, which was able to take advantage of existing Republican and kinship networks.[85]

[t] NRA members debate about which was more important—the supply networks or political mobilization. Kasfir suggests that the NRA accomplished both simultaneously. In their collections of food, the NRA also set up a network of supporters, which they then used to educate the populace about their political objectives and collect needed intelligence.[86]

effectiveness.[u] Numerous factors account for variation, including ideology, funding sources, and environmental factors. Ideology certainly can have an effect on whether or how insurgent groups pursue governance activities, but when groups are operating in a constrained environment, the mere professed desire to engage in governance activities is insufficient for execution of such governance. Funding constraints play a role in the relationship between insurgent groups and civilians. Funding from external sources, or from resources that can be extracted using little civilian labor, may contribute to more predatory relations between insurgents and civilians.[91]

Other factors related to the political and environmental context in which insurgents operate offer a more robust explanation of variation. One researcher developed a comprehensive framework for understating variation in comparative studies of shadow governments. Although limited space precludes a thorough discussion of each of the eight factors, several stand out in relation to the examples of shadow governments above. First is the extent of state penetration in areas in which insurgents emerge. Citizens used to a high level of security and social services from the state are more likely to expect and demand more of the same from insurgents seeking to supplant state rule. In Sri Lanka, for example, the high level of services provided by the state shaped the extent of government services offered by the LTTE. In competition with the state, rival militant groups, and nonviolent Tamils, the LTTE "went to great lengths to quell Tamil dissatisfaction with living conditions in rebel-controlled areas."[92] The RCD, by contrast, despite vociferous criticisms and demands for better governance from civilians, failed to successfully initiate an effective shadow government, relying instead on coercion and violence to enforce compliance.[93] Another factor shaping insurgent governance activities is the organization's political objectives. Those groups that have secessionist or ethno-nationalist claims exhibit more effort in securing legitimacy from their constituents through effective governance. While the RCD expected to quickly capture the capital, Kinhasa, and the associated state institutions, the LTTE sought to secure an independent state. Thus, the LTTE's governance activities were a concerted effort to prove its credentials to the population as a capable nation-state. For the RCD, by contrast, territorial control was mere "military expediency," and its efforts were limited to capturing, not creating, a nation-state.[94]

---

[u] For Mampilly, shadow governance effectiveness is the key variation in comparative studies, while Weinstein addresses levels of civilian participation in his comparative study of shadow governments.

PUBLIC COMPONENT

Today's insurgencies are more likely to include a public sphere in addition to the guerrillas, underground, and auxiliary common to insurgencies in preceding decades. The "public component" refers to the component of armed insurgencies that is public, or political, such as political parties, which often means that insurgent groups are simultaneously engaging in nonviolent and violent opposition. This section does not seek to uncover causal mechanisms to account for the inclusion of the public component but describes a variety of ways in which the public component, and its relationship to armed opposition groups, manifests.

Armed oppositions and political parties can have a variety of relationships.[v] In a study of linkages between terrorist groups and political parties, the most common relationship—evident in 57 of 203 total cases—was the formation of a terrorist group by existing political parties. In fact, among the variety of linkages uncovered by the research, in 134 of the 203 cases, the political party was typically the dominant element, or the progenitor, of the violent wing. This included cases where factions broke away from the political party to create a terrorist group and where the party supported an external terrorist group in addition to cases in which political parties outright created a violent wing.[96] It is important to note that this does not indicate that in all cases the political party was the dominating partner in the relationship. The reverse relationship, in which terrorist groups created a political wing, such as the IRA's creation of the party Sinn Féin, is much less common, occurring in only twenty-three cases. More promising than the mere numbers, however, is the quality of this relatively rarified transformation. Those political parties that turn to violence, like the Euskadi Ta Askatasuna (Basque Homeland and Freedom, or ETA), appear to vacillate between party politics and violence, while those terrorist groups that turn to politics, such as the PIRA, if the transition is a successful one, appear to be more stable.[97]

Insurgent groups can operate on dual tracks, engaging in armed opposition while simultaneously pursuing objectives through a political

---

[v] The research referred to in this paragraph establishes the spectrum of linkages between "terrorist" groups and political parties rather than "insurgent" groups. Leonard Weinberg included groups that met one or more of the following criteria: "1). the group was described as either 'terrorist' or 'urban guerilla;' 2). the group's activities included violent acts which (a) were perpetrated in some type of political context; (b) involved a symbolic or psychological effect, hence, aiming to influence a wider audience and not just harm the immediate victims, and (c) were aimed at non-combatants or civilians." In some instances, the terrorist groups included map onto the ARIS cases, which were included on the basis of criteria for insurgencies, such as the PIRA, Hizbollah, and the Palestine Liberation Organization (PLO)/HAMAS.[95]

wing. Of those that established political wings, many did so under conditions involving state repression and difficulties communicating with the public because of the clandestine nature of their operations. Of particular interest with regard to armed groups operating on dual tracks is the establishment of a political wing to facilitate a group's communication with its constituents and other audiences. Weinberg et al. note that terrorist groups, because of the nature of their activities, operate clandestinely, which makes it difficult for them to communicate their objectives and messages to a larger public. A "front" organization is critical in transmitting a group's plans for "political and social change to the public."[98] Within Lebanon, Hizbollah is able to "stage public rallies, make use of television and the World Wide Web and have its legislative representatives air its views in parliament." On the other hand, within Israel, Hizbollah must operate clandestinely. Such a front, or political wing, is also especially critical during negotiations with the government because it provides the government with a negotiating partner that is, ostensibly, not directly connected to the violence conducted by the military wing.

One of the quintessential examples of an organization with dual tracks is Hizbollah in Lebanon. Unlike many armed opposition groups, Hizbollah has doggedly laid infrastructure for political activities, including establishing schools, mosques, hospitals, and voluntary welfare associations. From early in its career, "the aim of the Hizbollah and Iran had been to strike roots in the Shiite society in Lebanon . . . ."[99] Hizbollah did not, however, contest a parliamentary election until 1992, several years after the death of Ayatollah Khomeini in 1989. Khomeini's death allowed Hizbollah leadership a bit of breathing room to follow a more independent path—one that included establishing a political party and participation in the Lebanese political process, a move now supported by Iran despite Hizbollah's abandonment of revolutionary goals. Hizbollah viewed participation as an opportunity to block any normalization of relations with Israel after the Gulf War, and it also provided the organization with an avenue of survival should it ever be induced to disarm. Additionally, the perks of political participation, including "access to political resources such as governmental posts, contracts, authorizations, permits, and public exposure" proved attractive as well. The election of Hassan Nasrallah to the post of Secretary General of the organization solidified its dual military and political trajectory.[100]

In addition to operating on dual tracks, insurgent groups can also wholly transition to the political process and cease armed struggle as a means to achieve their objectives. The factors that drive such a decision include ideological flexibility, strong centralized leadership, and internal cohesion among the support base. For the few groups that experienced spoiler violence, such as the Omagh bombing in 1998 by the Real IRA in protest of the PIRA's Good Friday Agreement, strong, committed leadership kept the momentum toward a peaceful conclusion to the conflict.[101] Much like Hizbollah, the PIRA and Sinn Féin operated on dual tracks for much of their campaign. Initially, Sinn Féin took a peripheral role to the military wing in the struggle. It was not until a generational change occurred within the leadership in the 1980s that Sinn Féin assumed a stronger role in the Republican struggle. The Provisionals originated as a radical splinter group from the Official IRA in 1969 after the latter's attempts to end abstentionism. As a result, in its early career, the PIRA was belligerently opposed to any participation in the political process because such participation would signal the legitimacy of the fundamentally unjust states of Northern Ireland and the Republic of Ireland. The watershed event that convinced the Provisionals of the instrumental value of the political process was the successful election of Bobby Sands to the British Parliament while on hunger strike in the Long Kesh Prison in 1981. His struggle and death on the sixty-sixth day of the strike mobilized popular support, both domestic and international, for the Provisionals. Recognizing the tactical value of the political process, the PIRA inched its way toward participation by occupying seats in the Northern Ireland local council elections. Gerry Adams took the reins of Sinn Féin in 1983, and by 1986, the Provisionals approved the pursuit of seats in the Republic of Ireland's parliament and lifted the longstanding ban on discussion of ending abstentionism. After a series of blundered operations that left civilians dead, and facing dwindling popular support, Sinn Féin and the Provisionals increasingly looked toward a political solution to the conflict, a solution facilitated by secret talks between Gerry Adams and the moderate leader of the Social Democratic and Labour Party (SDLP), John Hume.[102] Declaring a cease-fire in 1994, Sinn Féin was in uncharted territory but reportedly took comfort from the successful transition of the African National Congress (ANC). The peace process, which began in 1994, continued with fits and starts until the Good Friday Agreement was eventually reached in 1998. Unsurprisingly, the Republicans privy to the negotiations cited the issue of weapons decommissioning as the most difficult point of compromise. In 2005, the PIRA formally announced the end of its decades-long armed struggle and the decommissioning of its weapons.[103]

Insurgent conflicts in the post-Cold War world tend to conclude with an outcome other than the dramatic overthrow of the government by force of arms. As discussed above, the vast majority of conflicts today are intrastate as opposed to interstate. Those intrastate wars[w] are increasingly ending not in a decisive military victory but through a negotiated settlement.[x] A negotiated settlement is defined as "an ideal-type war termination in which neither side admits defeat and the combatants agree to end the violence and accept common terms on how to govern a postwar state," such as the Good Friday Agreement in 1998 that settled the Northern Ireland conflict.[106] A military victory, by contrast, is defined as "an ideal-type war termination in which one side explicitly acknowledges defeat and surrenders," such as the definitive defeat suffered by the LTTE in 2009.[107] In all civil wars ending between 1940 and 2000, 129 in total, most civil wars, 79, or 70 percent, ended in clear victory for one side. For the entire period, 22 wars, or 19 percent, ended in negotiated settlement. However, of the wars that ended in the 1990s, 41 percent ended through negotiated settlement, the same percentage of those ending in victory for one side. What is striking is that of all the wars that ended in negotiated settlement, two-thirds of those wars ended in the 1990s.[108] In a separate statistical analysis, researchers found that of the 38 civil wars they catalogued from 1945 to 1998, all but one included some provision for power sharing among the combatants.[109]

The net effect of the prevalence of negotiated settlements and power-sharing agreements is that more insurgents are "changing their stripes"[110] and demobilizing into the political process. Researchers find the inclusion of insurgent groups into the political process attractive along two primary lines. The first relates to the remediation of legitimate grievances, such as exclusion from political power, that spurred the onset of conflict. The second holds that inclusion remedies commitment problems in the peace process by providing insurgent groups with incentives for peace. Various actors, including international donors or concerned third parties, have favored inclusion of insurgent groups in several peace processes after the end of the Cold War. The entrance of insurgent groups into the political process occurred before the end of the Cold War, but most entered politics through the use of force, either

---

[w] Toft uses six criteria to define civil wars, and amalgamation of criteria from various respected scholars in the field. The criteria include a commonly used "death threshold," a macabre moniker for a criterion of at least an average of 1,000 battle deaths per year. This high death threshold excludes "smaller-scale" insurgencies, such as the Northern Ireland conflict. However, other conflict researchers, such as Nicholas Sambanis, use a death threshold of a total 1,000 battle deaths throughout the duration of the conflict, which would include Northern Ireland. As evidenced here, political scientists have struggled to agree upon a precise quantification of what constitutes a "civil war."[104]

[x] Negotiated settlements, and civil war termination in general, are thought to have increased in the post-Cold War environment for a number of reasons, including the withdrawal of U.S. and Soviet resources from proxy wars as well as increased pressure on the U.S. and the international community to intervene in civil wars.[105]

through victory over extant governments or former colonial powers.[111] Inclusion has led to several different results, including the emergence of former insurgencies as new government parties, such as the ANC in South Africa or Ḥarakat al-Muqāwamah al-'Islāmiyyah (Islamic Resistance Movement, or HAMAS) in the Palestinian territories; the emergence of former insurgencies as opposition parties, such as many Central American states; or, as in the case of UNITA in Angola, former insurgents taking up seats in parliament and the cabinet according to power-sharing mechanisms.[112]

Undergrounds and their associated insurgent groups face numerous challenges in this transformation from an illegal, armed opposition group to a bona fide actor in the political process, including those related to organizational, transitional justice, and security matters. The transition to conventional politics "requires adopting a new political culture, formulating a new programme, installing party organisational structures, recruiting party cadres, and building their capacity to govern."[113] Those insurgent groups that have operated on dual tracks, such as the PIRA and Sinn Féin, appear to adapt more readily to the changing environment but still face numerous obstacles. A leader of the ANC in South Africa notes that despite the organization's victory in the 1994 elections after the peace process, the ANC would have benefitted from paying more attention to building a team "ready to govern and build up its capacity to deliver."[114] Leaders within the Communist Party of Nepal–Maoist, or CPN-M, in Nepal anticipated obstacles to the transition:

> After 10 years of the People's War, we had entered into the phase of the peaceful development of the revolution. The form of our struggle had changed. Before, our activities were concentrated in rural areas and our main fighting forces were the PLA. But now, we had to do more in urban areas, with mass mobilisations and open activities as the primary focus of our work. We therefore had to train the party and PLA cadres in this new approach. For that purpose, Comrade Prachanda and I visited five regions throughout May and June 2006 to give political classes, mainly about how to develop the peaceful revolution.[115]

To address these issues, the CPN-M initiated many organizational changes to adapt the group to peaceful politics. PLA political commissars were transferred from the military wing to the new Party. Former members of the Central Committee became District-in-Charges responsible for dialoguing with other political parties. The CPN-M also shifted its organizational structures to match those of the state administration and dissolved Regional Bureaus previously used to facilitate communication between the Central Committee and cadres in favor of State Committees that were better reflections of the ethnic and geographic diversity of Nepal. Chairman Prachanda also dissolved all existing shadow government structures, the People's Governments, including the parallel judicial system, the People's Courts.[116]

In addition to obstacles in transforming political cultures and organizations, rebel groups entering the political process also face dilemmas of transitional justice, or how to "deal with the past." Transitional justice must navigate between society's need for justice for crimes committed during the course of the conflict and reconciliation of the warring parties and society. In South Africa, the interim constitution included provisions granting amnesty for offenses associated with political objectives during the conflict as well as establishing a Truth and Reconciliation Commission. Similar mechanisms calling for amnesty were established in other post-conflict settlements, such as in Aceh and Colombia.[117]

Lastly, insurgent groups face difficulties in decommissioning weapons and reintegrating fighters into a peaceful society. Duduoet's analysis of the decommissioning process in six case studies suggested that "the demobilisation and disarmament of insurgency movements can only flow out of the negotiation and democratisation or state reform processes." Combatants face security dilemmas during and after peace negotiations. In Colombia, 18 percent of M-19 was assassinated after the 1989 peace process. Unsurprisingly, premature demands for disarmament prior to addressing security concerns and substantive political issues can derail or hamper the peace process. Difficulties in the negotiations between CPN-M and the Nepal government arose primarily from the issue of disarmament when Nepal, strongly supported by the U.S. ambassador, demanded CPN-M's disarmament before it could enter the interim government. The parties eventually agreed to simultaneous and reciprocal cantonment. Final demobilization and integration of the PLA were not discussed until after elections and was negotiated by a "multi-party Army Integration Special Committee."[118] Settlements also rely on various tactics to integrate former combatants into society. The final agreement, reached in November 2011, included rehabilitation packages offering "formal and informal education, vocational training, agriculture and livestock training, and preparation for foreign training" for many PLA militants.[119]

THE POST-CONFLICT ENVIRONMENT

The peace after civil wars[y] is notoriously fragile—countries that have experienced a civil war have a high risk of relapsing into armed conflict once more. The Correlates of War (COW) database documents 108 civil wars from 1944 to 1997. Those 108 civil wars occurred in only 54 nations. Of those nations, only twenty-six experienced a single civil war, while ten had two civil wars, twelve had three, four had four, and two unlucky nations experienced five civil wars. Civil wars exact a high price in terms of human life—they lead to four times as many casualties as interstate wars during the same period.[120] For those who survive wars, health problems due to forced migration and poor preventative care further degrade their quality of life. The economic cost of civil war is also tremendously high. The typical civil war lasts seven years[121] at a hefty price tag of $3 billion a year; this cost is especially burdensome for the poorest countries in the world, where most civil wars occur.[122] It takes a country that has experienced a civil war an average of twenty-one years to regain its pre-war level of income.[123]

A host of factors put countries at risk for civil war—including poor economic development, ethnic fractionalization, weak states, and natural resources, among others[z]—but these same factors by themselves cannot adequately account for the frequency of civil war recurrence. A set of additional factors related to the civil war itself and to how the previous civil war ended have offered insight into why civil wars so frequently reignite. In relation to the former, civil wars that last longer are less likely to recur—this is the so-called "war weariness" theory. For each year a civil war lasts, the chances that hostilities will resume decrease by 10 percent. The deadliness of the conflict also affects the possibility of recurrence. The higher the casualties, the more hatred, hostility, and distrust among the combatants, making them more likely to resume violence.[124] Finally, the conditions under which the conflict terminated—e.g., a negotiated settlement or a decisive military victory—influence the potential for further conflict. Negotiated settlements, while preferable for humanitarian reasons, often carry the seeds of future violence.

As discussed above, civil wars are the most common type of conflict today. These changes in the type of conflict in the post-Cold War era

---

[y] It should be noted that many of the datasets social scientists use in their statistical analyses exclude smaller-scale insurgencies such as the Northern Ireland case. Also, as evidenced here, social scientists tend to describe intrastate conflicts as "civil wars" rather than insurgencies.

[z] With perhaps the exception of poor economic development, researchers debate the extent to which the remaining factors contribute to civil war onset.

are also matched by changes in how conflict is settled. Civil wars are increasingly likely to be settled not decisively on the battlefield but at the negotiating table. A negotiated settlement is a preferred method of settling political violence; the method is especially preferred by the international community, which often mediates settlements between combatants. However, negotiated settlements are more likely than military victories to lead to a resurgence of conflict. Civil wars ending in military victory were almost twice as likely to remain settled than those established by negotiated settlement or a cease-fire/settlement. From 1940 to 2000, ten out of seventy-nine, or 13 percent, of civil wars ending in military victory recurred; that number nearly doubled to five out of twenty-two, or 23 percent, in civil wars ending in negotiated settlement.[125] Disaggregating the military victory variable reveals more telling results. Military victories won by rebel groups are more likely to endure than those won by government forces. While eight out of forty-seven wars, or 17 percent, ending in government victory recurred, only 6 percent, or two out of thirty-two, wars ending in rebel victory recurred. It is worth noting here that government victories are still more likely to endure than negotiated settlements. Because more and more civil wars are ending in negotiated settlement, and because the international community, especially the United States, expresses a political preference for negotiated settlements for reasons discussed below, it is imperative to understand why negotiated settlements fail, what makes military victories, and in particular rebel victories, more secure, and the potential policy implications for these findings.

Negotiated settlements are attractive for numerous reasons. Settlements align with democratic norms prevailing in Western societies taking key roles in conflict intervention. Additionally, negotiated settlements are viable options for ending particularly protracted civil wars. Researchers have found that the longer a civil war endures, the less likely it is to be settled by a decisive military victory. In fact, it appears that the longer a civil war lasts, typically after the five-year mark, the more likely the conflict is to end through a negotiated settlement.[126, 127, aa] These findings suggest that negotiated settlements often arise out of conflicts that have reached a "mutually hurting stalemate" where each combatant acknowledges its inability to defeat the other militarily. Advocates of negotiated settlements argue that they save lives. However, the findings above, particularly in terms of the likelihood that civil wars will reignite after these settlements, suggest otherwise. Once a civil war reignites after a negotiated settlement, the conflict is deadlier than those

---

aa Mason and Fett found that those civil wars lasting longer than five years are more likely to be terminated in a negotiated settlement than a military victory, whether rebel or government.[128]

recurring after military victories. Although politically unpalatable, the prevalence of civil war recurrence after settlements has led some to support "giving war a chance," or adopting a policy of nonintervention at the least and, at most, supporting one combatant in a conflict to ensure a military victory.[129] Lastly, empirical evidence does not support the contention that negotiated settlements increase the quotient of democracy or economic prosperity in countries party to them. Although initial increases in democracy are apparent, those increases do not last past the initial election cycle typically mandated by the settlement.[130][ab]

Despite the difficulties facing countries in the post-conflict environment after a negotiated settlement, various measures have been found to alleviate civil war recurrence. Not all negotiated settlements are created equal. Those that endure are those that include power-sharing agreements among combatants across a wide spectrum of state power—politics, military, territory, and the economy. Each dimension added to the settlement increases the chance that the settlement will endure. Not surprisingly, the full implantation of military-sharing arrangements substantially reduces the risk of a return to hostilities. Only one in ten states that implemented military power-sharing returned to hostilities, while half of those states that failed to fully implement military power-sharing returned to war.[132] Demobilization and disarmament leave combatants especially vulnerable, leading to what researchers call a "credible commitment problem," where combatants lack assurance that their adversaries will follow the terms of a settlement. Introducing a third party that will step in to enforce the bargain, called a "third-party guarantee," is often successful in addressing this commitment problem, leading to a more stable peace.[133] Similarly, the introduction of peacekeeping forces is also positively associated with a more stable peace after a civil war, cutting the incidence of recurrence by half, in terms of both its potential military deterrence and the economic incentives it offers to rebel leaders and soldiers.[134]

Military victories, and in particular, rebel victories, are less likely to lead to a recurrence of hostilities than negotiated settlements. Studying how victorious rebels behave in the aftermath of a conflict can lend insight into maintaining peace. Most obvious is that a military victory indicates that the victor, whether rebel or government, has a monopoly on the legitimate use of force in the post-conflict environment:

[ab] Underscoring the importance of methodology in interpreting results of quantitative analysis, Mason et al. reached a different conclusion regarding the fragility of peace after negotiated settlements. Using a different dataset, they found that the peace after negotiated settlements was more fragile than government victory only initially but more durable in the long term. In their analysis, negotiated settlements lasting fourteen years or longer are less likely to result in a resurgence of conflict than those ended through a government victory.[131]

"Victorious armies are typically large, disciplined, and well equipped, and . . . more effective than the forces they defeated."[135] Violence perpetrated by losers in a military victory is criminal, but violence perpetrated by the victor is lawful and expected. After a negotiated settlement, however, both combatants retain the capacity to reignite the conflict.

Negotiated settlements offer many benefits, such as "provisions for development and reconstruction aid and . . . the redistribution of offices in postwar governments" to combatants, but promise little harm because security forces are typically fractured and anemic—all carrots and few sticks. By contrast, military victories often offer little in the way of benefits for the losing side, but, with a robust military, can promise harm to those defecting—many sticks, few carrots. As stated before, it is rebel victories that are least likely to return to conflict. What are the key differences between government and rebel victories? Increased repression usually follows government victories, with little to no reforms addressing the grievances underlying the conflict. By contrast, rebel governments, in an effort to consolidate their power and legitimize their rule, tend to enact far-reaching reforms.[136] Conflict is more likely to recur after government victories because while governments repress rebellion, they do not fully annihilate rebel groups or address grievances, leading to a shorter duration of peace before conflict breaks out again after rebels "regroup, rebuild, and resume combat."[137] In this regard, rebel victories are better at eliminating "multiple sovereignty"—when one or more organized armed challengers that command significant popular support exist in a state.[138,ac] A salutary mixture of benefits and harm—carrots and sticks—appears to be the most effective deterrent to civil war recurrence.

If a security sector capable of deterring a reignition of hostilities is part and parcel of an enduring peace, then negotiated settlements are often lacking. Many negotiated settlements take great pains to institute carefully crafted political institutions and disarmament, demobilization, and reintegration (DDR) programs, but most are lacking in provisions for comprehensive security sector reform (SSR).[140,ad] SSR can include the restoration of order, the neutralization of nonstate militias, the rebuilding of security forces to prepare them for the maintenance of public order, and the development of institutions to monitor and support security forces. Where provisions exist, resources

---

ac Rebels dismantle multiple sovereignty more effectively than government victory if they can remain in power for about four years, usually the time it takes to consolidate victory and defeat remnants of incumbent government and rival rebel groups.[139]

ad The security sector comprises the institutions "that have the authority to order the threat of force or use force to protect state and civilians," typically domestic police, military, and, sometimes, criminal justice sectors.[141]

allocated to the implementation of SSR are usually meager. Failure to attend to SSR can have "devastating consequences," leaving an array of militaries, militias, or rebels that can compromise the prospect of peace. Including comprehensive SSR in negotiated settlement provisions fosters the "mutual harm, mutual benefit" positively associated with a stable peace.[142]

ENDNOTES

1   Andrew R. Molnar, William A. Lybrand, Lorna Hahn, James L. Kirkman, and Peter B. Riddleberger, *Undergrounds in Insurgent, Revolutionary, and Resistance Warfare* (Washington, DC: Special Operations Research Officer, The American University, 1963).

2   Lotta Harbom and Peter Wallensteen, "Armed Conflicts, 1946–2009," *Journal of Peace Research* 47, no. 4 (2010): 503.

3   Lotta Harbom, Erik Melander, and Peter Wallensteen, "Dyadic Dimensions of Armed Conflict, 1946–2007," *Journal of Peace Research* 45, no. 5 (2008): 700.

4   Ian Spears, "States-within-States: An Introduction to their Empirical Attributes," in *States-within-States: Incipient Political Entities in the Post Cold War Era*, ed. Paul Kingston and Ian Spears (New York: Palgrave Macmillan, 2004), 28.

5   Zachariah Cherian Mampilly, *Rebel Rulers: Insurgent Governance and Civilian Life during War* (Ithaca, NY: Cornell University Press, 2011), 61.

6   Nelson Kasfir, "Guerrillas and Civilian Participation: The National Resistance Army in Uganda, 1981–86," *Journal of Modern African Studies* 43, no. 2 (2005): 86.

7   Alan Cowell as quoted in Spears, *States-within-States*, 21.

8   Ibid., 20.

9   Chuck Crossett and Summer Newton, "The Provisional Irish Republican Army: 1969–2001," in *Casebook on Insurgency and Revolutionary Warfare, Volume II: 1962–2009*, ed. Chuck Crossett (Alexandria, VA: U.S. Army Publications Directorate, in press).

10   Timothy P. Wickham-Crowley, "The Rise (and Sometimes Fall) of Guerrilla Governments in Latin America," *Sociological Forum* 2, no. 3 (1987): 482.

11   Harbom, Melander, and Wallensteen, "Dyadic Dimensions of Armed Conflict," 702.

12   Harbom and Wallensteen, "Armed Conflicts," 501.

13   Harbom, Melander, and Wallensteen, "Dyadic Dimensions of Armed Conflict," 704.

14   Ibid.

15   Newton, "The Provisional Irish Republican Army (PIRA): 1969–2001."

16   Kasfir, "Guerrillas and Civilian Participation," 279–280.

17   Spears, *States-within-States*, 15–34.

18   Ibid., 22–23.

19   Tony Addison and Syed Mansoob Murshed, *The Fiscal Dimensions of Conflict and Reconstruction* (Helsinki: United Nations University, World Institute for Development Economics Research, 2001), 1–2.

20   Ibid., 5

21   Spears, *States-within-States*, 25.

22   James B. Love, *Hezbollah: Social Services as a Source of Power* (Hurlbert Field, FL: JSOU Press, 2010), 1.

23   Spears, *States-within-States*, 26.

[24] Love, *Hezbollah*, 21.

[25] Ibid., 26.

[26] Ibid.

[27] James N. Rosenau, "Patterned Chaos in Global Life: Structure and Process in the Two Worlds of World Politics," *International Political Science Review* 9, no. 4 (1988): 2–5.

[28] Nelson Kasfir, *Dilemmas of Popular Support in Guerrilla War: The National Resistance Army in Uganda, 1981–86* (Hanover, NH: Dartmouth College, 2002), http://www.yale.edu/macmillan/ocvprogram/licep/6/kasfir/kasfir.pdf.

[29] Rosenau, "Patterned Chaos in Global Life," 2–5.

[30] Paul Kingston, "States-within-States: Historical and Theoretical Perspectives," in *States-within-States: Incipient Political Entities in the Post Cold War Era*, ed. Paul Kingston and Ian Spears (New York: Palgrave McMillan, 2004), 3.

[31] Ken Menkhaus, "Governance without Government in Somalia: Spoilers, State Building, and the Politics of Coping," *International Security* 31, no. 3 (2007): 85.

[32] James Forest, "Engaging Non-State Actors in Zones of Competing Governance," *Journal of Threat Convergence* 1, no. 1 (2010): 10.

[33] Spears, *States-within-States*, 20.

[34] Mark Duffield, "Globalization of War Economics: Promoting Order or the Return of History?" *Fletcher Forum of World Affairs* 23, no. 2 (1999): 22.

[35] Steven Metz, *Rethinking Insurgency* (Fayetteville, AR: Juniper Grove, 2007), 11.

[36] Dani Rodrik, "How Far Will International Economic Integration Go?" *The Journal of Economic Perspectives* 14, no. 1 (2000): 182.

[37] Metz, *Rethinking Insurgency*, 11–12.

[38] Ibid.

[39] Duffield, "Globalization of War Economies," 22.

[40] Mampilly, *Rebel Rulers*, 43.

[41] Duffield, "Globalization of War Economies," 21. Duffield calls this condition "durable disorder." It is important to note that not all scholars reach these conclusions. Some, for instance, argue that the prevalence of weak states simply means that the state as a form of political organization is on its way out, to be replaced, perhaps, with something resembling world governing structures. See Susan Strange, "The Westfailure System," *Review of International Studies* 25, no. 3 (1999): 345–354; and Susan Strange, "The Erosion of the State," *Current History: A Journal of Contemporary World Affairs* 96, no. 613 (1997): 365–369.

[42] Kingston, "States-within-States," 8.

[43] Mampilly, *Rebel Rulers*, 152.

[44] Field Manual 3-24, *Counterinsurgency* (Washington, DC: Headquarters, Department of the Army, 2006), 1–65.

[45] Paul Collier and Anke Hoeffler, *Greed and Grievance in Civil War* (Washington, DC: World Bank, Development Research Group, 2000), 1–2.

[46] Paivi Lujala, "The Spoils of Nature: Armed Civil Conflict and Rebel Access to Natural Resources," *Journal of Peace Research* 47, no. 1 (2010): 15–28. A competing explanation is that natural resources do not encourage insurgents to fight, leading to an increased risk of conflict for states with natural resources, but that abundant resources result in poor government policy choices and a weak state. See James D. Fearon and David D. Laitin, "Ethnicity, Insurgency, and Civil War," *American Political Science Review* 97, no. 1 (2003): 75–90.

[47] Mary Kaldor, *New & Old Wars: Organized Violence in a Global Era* (Stanford, CA: Stanford University Press, 2007).

[48] Duffield, "Globalization of War Economies," 27.

[49] Metz, *Rethinking Insurgency*, 30.

[50] Assis Malaquias, "Making War & Lots of Money: The Political Economy of Protracted Conflict in Angola," *Review of African Political Economy* 28, no. 90 (2001): 523.

[51] Ibid.

[52] Ibid., 531.

[53] United Nations Security Council, *Report on the Panel of Experts on Violations of Security Council Sanctions Against UNITA* (2000).

[54] Ibid.

[55] Ibid.

[56] Duffield, "Globalization of War Economies," 27. For a time, the protracted conflict in Angola was aided by de Beer's "no-questions asked" policy of diamond buying that allowed groups like UNITA to finance their operations and accumulate wealth. A United Nations report on the subject also indicates that the lax regulations in Antwerp, the world's premier diamond-buying market, enacted to facilitate trade also aided the organization. Since the report was published in 2000, de Beer's has opted to cease buying diamonds from Angola in an attempt to halt the purchase of UNITA's blood diamonds, which has met with a measure of success and driven down the prices UNITA can expect for its rough diamonds. United Nations Security Council, *Report on the Panel of Experts on Violations of Security Council Sanctions Against UNITA*, para. 89 and 96.

[57] Mampilly, *Rebel Rulers*, 50.

[58] Wickham-Crowley, "The Rise (and Sometimes Fall) of Guerrilla Governments in Latin America," 473–499.

[59] Ibid., 481.

[60] Ibid., 482.

[61] Mampilly, *Rebel Rulers*.

[62] Clifford Geertz, *Negara: The Theatre State in Nineteenth-Century Bali* (Princeton, NJ: Princeton University Press, 1980).

[63] Mampilly, *Rebel Rulers*, 56–57.

[64] Ibid., 115.

[65] Ibid., 118–119.

[66] Ibid., 119–120.

[67] Ibid., 120–123.

[68] Ibid., 123–127.

[69] Ibid.

[70] Ibid., 182.

[71] Ibid., 190–208.

[72] Ibid.

[73] Ibid., 191.

[74] FM 3-24, 1–65.

[75] Kasfir, "Guerrillas and Civilian Participation," 272.

[76] Ibid.

[77] Ibid.

[78] Ibid., 273.

[79] Andrew Silke, "Rebel's Dilemma: The Changing Relationship between the IRA, Sinn Fein and Paramilitary Vigilantism in Northern Ireland," *Terrorism and Political Violence* 11, no. 1 (1999): 75.

[80] Ibid., 81.

[81] Ibid., 71.

[82] Ibid., 84.

[83] Jeremy M. Weinstein, *Inside Rebellion: The Politics of Insurgent Violence* (Cambridge; New York: Cambridge University Press, 2007), 176–177.

[84] Kasfir, "Guerrillas and Civilian Participation," 272.

[85] Ibid., 282.

[86] Ibid., 284.

[87] Ibid.

[88] Ibid., 283.

[89] Ibid., 284–285.

[90] Weinstein, *Inside Rebellion*, 178–180.

[91] Ibid., 195–197.

[92] Mampilly, *Rebel Rulers*, 211.

[93] Ibid., 212.

[94] Ibid., 214–217.

[95] Leonard Weinberg, Ami Pedahzur, and Arie Perliger, *Political Parties and Terrorist Groups* (London; New York: Routledge, 2009).

[96] Ibid.

[97] Ibid.

[98] Ibid.

[99] Ibid.

[100] Ibid.

[101] Veronique Dudouet, *From War to Politics: Resistance/Liberation Movements in Transition* (Berlin: Berghof-Forschungszentrum für Konstruktive Konfliktbearbeitung, 2009), 47–48.

[102] Newton, "The Provisional Irish Republican Army (PIRA): 1969–2001."

[103] Bairbre De Brun, *The Road to Peace in Ireland* (Berlin: Berghof-Forschungszentrum für Konstruktive Konfliktbearbeitung, 2008), 14–15.

[104] Michael W. Doyle and Nicholas Sambanis, "International Peacebuilding: A Theoretical and Quantitative Analysis," *American Political Science Review* 94, no. 4 (2000): 779–801.

[105] Monica Duffy Toft, "Ending Civil Wars: A Case for Rebel Victory?" *International Security* 34, no. 4 (2010): 14–15.

[106] Ibid., 11.

[107] Ibid.

[108] Ibid., 12–14.

[109] Matthew Hoddie, "Power Sharing in Peace Settlements: Initiating Transition from Civil Wars," in *Sustainable Peace: Power and Democracy After Civil Wars*, ed. Philip G. Roeder and Donald Rothchild (Ithaca, NY: Cornell University Press, 2005), 83, 85. Civil wars are again defined here as resulting in an average of 1,000 battle deaths each year. The exception was a failed agreement to end the civil war in Angola. Hoddie and Hartzell define power sharing in terms of both formal and informal institutions and policies.

[110] Mimmi Soderberg Kovacs, "When Rebels Change Their Stripes: Armed Insurgents in Post-War Politics," in *From War to Democracy: Dilemmas of Peacebuilding*, ed. Anna K. Jarstad and Timothy D. Sisk (New York: Cambridge University Press, 2008).

[111] Hoddie, "Power Sharing in Peace Settlements," 138, fn.1.

[112] Kovacs, "When Rebels Change Their Stripes," 138–139. UNITA eventually went back to armed struggle, however.

[113] Dudouet, *From War to Politics*, 39.

[114] Ibid.

[115] As quoted in Kiyoko Ogura, *Seeking State Power: The Communist Party of Nepal (Maoist)* (Berlin: Berghof-Forschungszentrum für Konstruktive Konfliktbearbeitung, 2008), 41.

[116] Ibid., 41–42.

[117] Dudouet, *From War to Politics*, 39.

[118] Ibid., 39–42.

[119] Prasant Jha, "Maoist Combatants Reject Rehabilitation Option," *The Hindu*, November 21, 2011.

[120] T. David Mason, *Sustaining the Peace After Civil War* (Carlisle Barracks, PA: Strategic Studies Institute, U.S. Army War College, 2007), 2.

[121] Paul Collier, Anke Hoeffler, and Måns Söderbom, "On the Duration of Civil War," *Journal of Peace Research* 41, no. 3 (2004): 253–273.

[122] Paul Collier, Anke Hoeffler, and Mans Soderborn, "Conflicts," in *Global Crises, Global Solutions*, ed. Bjorn Lomborg (Cambridge; New York: Cambridge University Press, 2004), 129–156.

[123] Ibid.

[124] T. David Mason, Jason Michael Quinn, Mehmet Gurses, and Patrick T. Brandt, "When Civil Wars Recur: Conditions for Durable Peace After Civil Wars," *International Studies Perspectives* 12, no. 2 (2011): 171–189.

[125] Monica Duffy Toft, *Securing the Peace: The Durable Settlement of Civil Wars* (Princeton, NJ: Princeton University Press, 2010).

[126] James D. Fearon, "Why Do Some Civil Wars Last So Much Longer Than Others?" *Journal of Peace Research* 41, no. 3 (2004): 275–301.

[127] T. David Mason, Joseph P. Weingarten, and Patrick J. Fett, "Win, Lose, or Draw: Predicting the Outcome of Civil Wars," *Political Research Quarterly* 52, no. 2 (1999): 239–268.

[128] T. David Mason and Patrick J. Fett, "How Civil Wars End: A Rational Choice Approach," *Journal of Conflict Resolution* 40, no. 4 (1996): 546–568.

[129] Edward N. Luttwak, "Give War a Chance," *Foreign Affairs* 78, no. 4 (1999): 36–44. Choosing the "good" combatant over the "bad" combatant contains its own set of problems.

[130] Toft, *Securing the Peace: The Durable Settlement of Civil Wars*.

[131] Mason et al., "When Civil Wars Recur: Conditions for Durable Peace After Civil Wars," 184–185.

[132] Caroline A. Hartzell and Matthew Hoddie, *Crafting Peace: Power-Sharing Institutions and the Negotiated Settlement of Civil Wars* (University Park, PA: Pennsylvania State University Press, 2007).

[133] Barbara F. Walter, *Committing to Peace: The Successful Settlement of Civil Wars* (Princeton, NJ: Princeton University Press, 2002). Walter also notes that it is not coming to an agreement that is the most difficult component of negotiations, even on issues with so-called "indivisible stakes" like identity, but the above issue of profound insecurity accompanying dismantling the capacity for organized violence by combatants party to negotiations.

[134] Virginia Page Fortna, *Does Peacekeeping Work?: Shaping Belligerents' Choices After Civil War* (Princeton, NJ: Princeton University Press, 2008). Military deterrence forms part of the success of peacekeeping missions. Peacekeeping missions also offer economic incentives to the peacekept, including government positions for rebel leaders and compensation to rebel soldiers as part of a demobilization program. See also Doyle and Sambanis, *International Peacebuilding: A Theoretical and Quantitative Analysis*, 779–801; Mason et al., "When Civil Wars Recur: Conditions for Durable Peace After Civil Wars," 171–189.

[135] Toft, *Securing the Peace: The Durable Settlement of Civil Wars*.

[136] Ibid. In the aftermath of a rebel victory, governments tend to exhibit more democratic tendencies than those of states after a government victory. However, most rebel governments are still authoritarian, not bona fide democracies.

[137] Mason et al., "When Civil Wars Recur: Conditions for Durable Peace After Civil Wars," 185.

[138] Ibid., 172.

[139] Ibid., 185.

[140] Toft, "Ending Civil Wars: A Case for Rebel Victory?," 32.

[141] Toft, *Securing the Peace: The Durable Settlement of Civil Wars.*

[142] Ibid. Toft includes case studies of two successful negotiated settlements, one in Uganda and other in El Salvador, that did include SSR, connecting the relatively stable peace in both countries with these provisions.

CHAPTER 10.

SUBVERSION AND SABOTAGE

CHAPTER CONTENTS

Jerome M. Conley

INTRODUCTION

An insurgency is formally defined by the U.S. Department of Defense as "the organized use of subversion and violence by a group or movement that seeks to overthrow or force change of a governing authority."[1] Within this definition, the concept of "armed conflict" is fairly well understood and clearly falls within the scope of guerrilla warfare. But the full spectrum of activities that could be considered "subversive" and the viable tools within a strategy for overthrowing a government are somewhat less clear and can cover a broad range of kinetic and nonkinetic tactics. This chapter discusses the objectives and tactics associated with "subversion" and includes a detailed exploration of "sabotage" as a key tool for subversion. Included within this discussion is the emergence of nonkinetic sabotage capabilities, such as cyber attacks, that can generate destructive results without the use of explosives and other traditional means of sabotage.

SUBVERSION

As depicted in the diagram in Figure 10-1, subversive activities are the purview of the underground component of an insurgency but stop short of full armed conflict and guerilla warfare. Encompassing a broad scope of tactics and objectives, subversion is defined as the "actions designed to undermine the military, economic, psychological, or political strength or morale of a governing authority."[2] In this regard, the list of potential subversive activities is quite broad and can include propaganda and information operations all the way through selective sabotage and the targeted killing of key individuals in a sitting government.

171

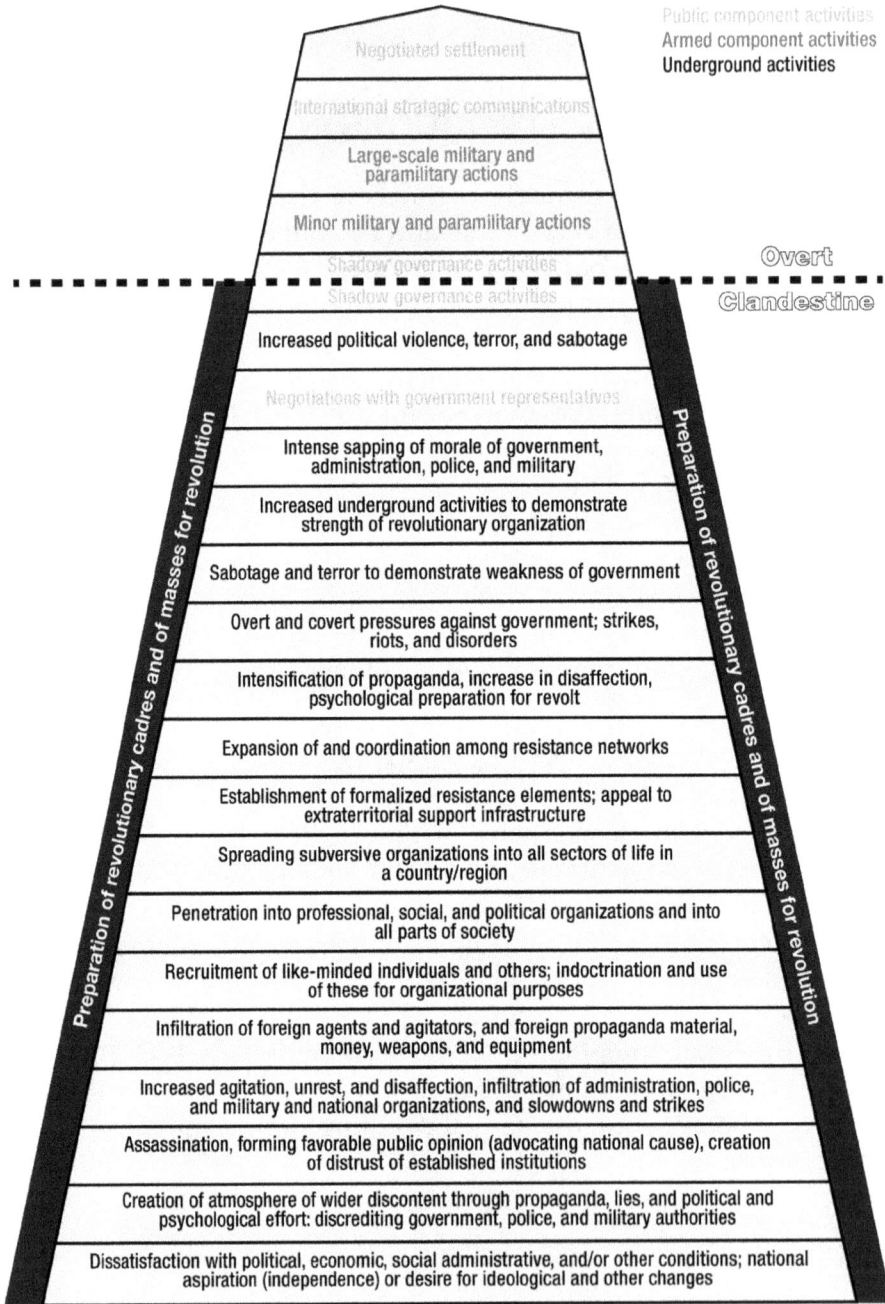

Figure 10-1. Underground operations.

Objectives of Subversion

To achieve the goal of undermining a governing authority, insurgency planners may employ several approaches during the latent, build-up phase of the insurgency in order to generate support for the movement while simultaneously degrading the influence and power of the ruling government. The primary audience for these subversive activities is the domestic population, although subversive efforts are often extended internationally as well in order to pull in political, economic, and even military support from abroad. Subversion objectives include:

- Undermining the authority and reputation of the military and security services
- Degradation of the economic strength of the country and/or the sitting government
- Infiltration and subversion of key organizations within the sitting government
- Psychological operations
- Undermining the political authority of the sitting government

Undermine Military and Security Services

A primary strategy undertaken by insurgent forces is to seek to undermine the authority and influence of the opposing military and security services as a means of increasing support for the insurgency. In many of these cases, there is already a history of abuse by the ruling government, so the process of documenting and highlighting these abuses may provide sufficient evidence for degrading the authority of these security and military services. The New People's Army (NPA) in the Philippines effectively leveraged the abuses by the Marcos regime to boost recruitment and mobilize new members. Abuses by the Colombian security forces were a common theme in public statements by the Fuerzas Armadas Revolucionarias de Colombia (Revolutionary Armed Forces of Colombia, or FARC), as were references by the Frente Farabundo Martí para la Liberación Nacional (Farabundo Marti National Liberation Front, or FMLN) to government-led death squads in El Salvador. Interviews with militants and nongovernmental organization (NGO) workers in Sri Lanka identified abuses by the Sinhalese security forces as the primary motivating factor behind them joining the Liberation Tigers of Tamil Eelam (LTTE). And the Ushtia Çlirimtare e Kosovës (Kosovo Liberation Army, or KLA) in Kosovo achieved tremendous success in generating international support—and NATO military involvement—with its persistent messaging about Serbian human rights abuses.

173

In the absence of abuse, or in situations where abuse by security or military organizations has not recently occurred, insurgent groups may spread false reports about new abuses; attempt to instigate new abuses against the population by the security services; or, in some situations, conducted abuses themselves that can then be blamed on the military or security apparatus. The 1972 "Bloody Sunday" shootings of twenty-six unarmed Catholic protesters by British Army personnel completely undermined the authority of the British Army as a neutral protection force in Northern Ireland and served to generate tremendous additional support for the Provisional Irish Republican Army (PIRA). However, despite two investigations by the British government into the events leading up to the shootings, debate still lingers over the degree to which PIRA snipers and youths throwing rocks and Molotov cocktails were intentionally used by the PIRA to instigate a response by the British Army.[a]

In an even more extreme scenario involving an insurgent group conducting atrocities against the population and then blaming these actions on the governing military force, the Revolutionary United Front (RUF) rebels in Sierra Leone performed "false flag operations" in which they wore Sierra Leone Army uniforms during some of their raids and attacks on villages, thus undermining an already waning public trust in the army that was supposed to protect them.[3]

Economic Degradation

The history of undergrounds and the use of subversion typically include the deliberate degradation of the economic capabilities of a country and/or sitting government in order to reduce its financial stability while simultaneously creating hardships on the population with associated backlashes against the government. The means for achieving an economic degradation can vary in scope and depth. During the German occupation of Poland in World War II, members of a Polish underground organization shut down factories for one day by sending forged orders to factories and workshop managers that proclaimed

---

[a] The first investigation (the "Widgery Tribunal") was conducted in 1972, and the second investigation (the "Saville Inquiry") was completed in 2010. Both investigations found that the shootings were unjustified, with the Saville Inquiry being much more critical in condemning the actions taken by the soldiers. The Saville Inquiry found, however, that PIRA leaders were present during the protest march and at least one was armed with a Thompson submachine gun, but it could find no conclusive evidence that the soldiers felt threatened by snipers or petrol bombs. In response to the release of this final report, a paratrooper who was in Derry during the shootings disagreed with the perspective that the soldiers did not feel threatened. See http://www.nationalarchives.gov.uk/webarchive/public-inquiries-inquests-bsi.htm and http://www.belfasttelegraph.co.uk/opinion/bloody-sunday-inquiry--a-soldiers-view--i-was-in-derry-that-day-i-just-wish-the-army-hadnt-been-14843878.html.

May 1 as "Nazi Labor Day." Written on the letterhead of the German Labor Bureau and delivered just before May 1, the orders used standard Nazi terminology and stated that all workers were to have twenty-four hours of paid leave. As a result, most production in Poland ceased for a day, including that at the important Ursus Tank Works and the gigantic railway repair installations at Pruszkow. Reportedly, the production losses were comparable with those resulting from a minor attack by the British Royal Air Force (RAF).

In Nigeria, where the country's economy is heavily reliant on oil revenues, the Movement for the Emancipation of the Niger Delta (MEND) periodically targeted the oil industry in order to force the Nigerian government to give in to its demands. In 2004, "Operation Locust Feast" was initiated by a precursor militant group to the MEND to target pipelines and oil workers and resulted in global oil prices going above $50/barrel for the first time. Upon the formal creation of MEND in early 2006, it immediately conducted another armed campaign against the oil industry in the Niger Delta, which resulted in a 25 percent reduction in oil output from the Niger Delta region.

In Sierra Leone, the RUF determined that a sharp reduction in the harvest yield from the country's farmers would cause significant economic hardship for the government and the population. During the harvest season, therefore, the RUF attacked farming villages and amputated the hands or arms of field workers as a warning to others that the crops should not be harvested, thus denying the government this critical food supply.

Removal of Dangerous Persons

During the early, latent phase of an insurgency, members of an underground may determine that certain people in the government or society pose a potential risk to the insurgency because of their role in active countermeasures or the perception that they are (or will) undermining the ability of the insurgency to mobilize the population. In these situations, they may decide to remove the person through kidnapping, intimidation, or murder. The Polish underground again demonstrated its creativity during World War II by having Volksdeutschen (citizens of German descent and sympathies) transferred onto active duty in the German army by forging a letter to Berlin with their signatures requesting the honor to serve in the military. The PIRA adopted a more visible approach to dealing with sympathizers and informants by "kneecapping" these individuals so their permanent injury (and limp) would serve as a reminder to other potential informants. In Sierra Leone, the RUF would decapitate a village elder when entering a new operating area and place the head on a stake outside of town as an immediate warning and reminder to those who might pass intelligence on to the government forces. And in Sri Lanka, the LTTE tried to avoid challenges from the government as well as other Tamil separatist movements by killing thirty-seven politicians and police informants, twenty-four of whom were fellow Tamils. In Northern Nigeria, the Muslim extremist organization Boko Haram routinely targeted moderate Muslim clerics who were viewed as espousing a more moderate version of Islam that might undermine the jihadist message of Boko Haram.

Undermine Political Authority and Morale

A central approach for undermining an opposition government is to instigate actions that undermine the political authority of this government. In 1992, Kosovar Albanians elevated the level of their opposition to Serbian rule over the region of Kosovo by holding their own presidential elections, which were declared illegal by Belgrade. The Kosovars also set up a parallel system of government, schools, clinics, and tax collection. In Ukraine, the Orange Revolution (November 2004–January 2005) provided a unique example of an insurgency won completely during the latent phase of the movement without a shot being fired. Conducted within the constraints of the Ukrainian political process, this revolution leveraged exit polls, social media, mass media, and the Internet to educate the population about election rigging and to mobilize the population to action. During the 1996 national elections in Sierra Leone, the RUF again demonstrated its brutality when it conducted "Operation Stop Elections," which involved amputating the hands of potential voters as a counter-message to the ruling party's campaign slogan that "The future is in your hands."

Organizational Subversion

The ability of an underground to influence or control key organizations in a country can shape the strategy and outcome of an insurgency in that country. In El Salvador and Northern Ireland, where the Catholic Church is very influential, the ability to garner overt support from priests and nuns provided a degree of legitimacy to the insurgencies that were taking place. In Poland, the development of an insurgency within the structure of major labor unions enabled the Solidarity movement to effectively control the keys to the nation's economy and slowly negotiate for more political freedoms and the eventual ascendancy of Lech Walesa to the presidency.

As stated earlier in the chapter on recruiting, subversion carries the meaning of corrupting someone's morals or loyalties and is a key tool in taking aim at those within the government—i.e., officials, administrators, government workers, police, or military personnel. The Viet

Cong doctrine of *binh van* (promotion of desertion and defection from the government of Vietnam) is especially relevant in that it included the use of agitation, persuasion, coercion, and threats to get key officials, both military and civilian, to weaken the government's ability to rule as well as to swell the ranks of the insurgency.

If an underground is unable to infiltrate or influence existing organizations, it can create "front companies" or organizations that serve as an innocent facade to its actual work. These organizations will usually espouse some worthy cause that will engender the support of respectable members of the community, at least to the extent of permitting the use of their names, but the leadership of the organization is kept firmly in the hands of underground members.

SABOTAGE

The term "sabotage" is defined by the U.S. military as "An act or acts with intent to injure, interfere with, or obstruct the national defense of a country by willfully injuring or destroying, or attempting to injure or destroy, any national defense or war materiel, premises, or utilities, to include human and natural resources."[4] Sabotage can include both kinetic and nonkinetic activities. Operatives can plan and conduct both general and selective sabotage.

In selective sabotage, the underground tries to incapacitate installations that cannot be easily replaced or repaired in time to meet the government's crucial needs. Tactical targets, such as a bridge essential for transporting troops and supplies to a battle area, are concentrated on because strategic targets, such as factories, are much harder to incapacitate and must be incapacitated for a much longer period. Selective sabotage can also include a deliberate, nonkinetic strike on a specific information node or piece of critical infrastructure, such as a cyber attack on a supervisory control and data acquisition (SCADA) system or transportation center. An underground may also undertake sabotage not only to hamper the government's military effort but also to encourage the populace to engage in general acts of destruction. Such acts serve as a form of propaganda and commit people more firmly to the cause.

Undergrounds generally use simple explosives because their members usually are not trained demolition experts and the information and materials for developing improvised explosives are both easily acquired. Plastic explosives are ideal because they are easily stored, simple to use, and can provide significant and predictable levels of damage. Members receive training through manuals, directives, clandestine newspapers, leaflets, radio broadcasts, or personal instruction from military units. In some cases, insurgent groups may receive training from other insurgent groups, such as the well-documented case of PIRA members traveling to Colombia to train the FARC in improvised explosives. To foster general sabotage, the underground often instructs the population in the use of such simple devices as Molotov cocktails, tin-can grenades, and miscellaneous devices for causing fire or damage to small equipment. In preparation for a significant protest march in Derry, Northern Ireland, in 1969, the IRA lined up "tens of thousands" of empty milk bottles to be used for petrol bombs. Within the United States, the "Anarchist Cookbook" became a well-known publication among antigovernment and white supremacist groups for its detailed descriptions on making improvised explosives.

Selective Sabotage

A principal advantage of underground sabotage is that it can succeed in destroying or disabling a target not easily reached by conventional means. An insurgent leader in France during World War II said that this consideration often governed British action against targets in France. Launching an RAF attack on important targets in France required aircraft and trained flight crews whose main task was to hit targets in Germany or Italy. The chance of successfully hitting a small target was very low, even with the refinements of precision bombing. One saboteur could—with more personal risk, but also with far less expenditure of total lives and money—obtain more certain results in a shorter time.

Timing, of course, is critical. To blow up a bridge without considering the immediate needs of the enemy might only put an unnecessary hardship on the populace, but to destroy it at a time when it is vital to the enemy's troop and supply movements would be of real tactical value. Undergrounds typically excel at selective sabotage precisely because they can fine-tune the timing of the destruction. Thus they can coordinate an act of sabotage with operations by a friendly military force, or they can time an attack so as to disrupt a critical operation by the enemy.

The FMLN in El Salvador achieved significant gains by blowing up tens of bridges, including the two most important ones, effectively cutting the country in two by destroying all the main bridges that crossed the Rio Lempa, the economic lifeline that was the U.S. equivalent of the Mississippi River. The FMLN also destroyed dozens of coffee, sugarcane, and cotton installations, including all of the largest and most important ones; and it continuously interrupted the electrical system in more than 80 percent of the country. By collaborating with local workers from the oil industry and exploiting the difficulty in protecting hundreds of miles of pipelines, MEND was repeatedly successful in shutting down sections of the Niger Delta's oil infrastructure.

In Madrid (March 11, 2004), London (July 7, 2005), and Moscow (March 29, 2010), bombings were carried out on the metro systems in those capitals, killing 191 people in Spain (1,800 wounded), 52 in England (700 wounded), and 40 in Russia (100 wounded) while shutting down the critical transportation infrastructure of those cities. Carried out by independent cells that appeared to be sympathetic to Al Qaeda ideology (Madrid and London) and the Chechen separatist movement (Moscow), the bombings in two of the cities (Madrid and Moscow) were also suicide attacks.

Tactical Aspects

A number of explosives can be used in sabotage, but only a few can be handled easily by underground members. These persons generally are not demolition experts and therefore require explosives that are relatively safe and easy to use. Nitroglycerine is potent but unstable; almost any jarring can cause it to explode. When mixed with sawdust or other absorbent material, making dynamite, it is safer but still very sensitive in warm temperatures. More suitable for use by undergrounds is trinitrotoluene, or TNT, which is so stable that even a piercing bullet might not cause an explosion. Setting it off requires an embedded blasting cap of gunpowder. An improved explosive can be made by mixing TNT with hexogene; the product is a malleable but equally powerful explosive. Although it is sometimes called cyclonite, or RDX, it is often referred to as plastic explosive. This mixture is ideal for underground use because not only is it stable—it can be stamped on, cut, frozen, or fried, and it will not explode—but it can also be molded for any use and readily stuck to surfaces, as putty. Like TNT, it must be detonated by a blasting cap, and also like TNT, a one-half pound charge of "plastic" can kill or severely wound a person standing a few feet away.

For insurgents, therefore, the ability to establish a supply line and access to plastic explosives can be quite critical to the ability to conduct an effective campaign. For the PIRA, that supply chain included sources in the United States but also Colonel Muammar Gaddafi in Libya, who sent a shipment of weapons and explosives to the PIRA in 1987 that was interdicted by French authorities at sea and that included over 150 tons of armaments: 1,000 AK-47s, 1 million rounds of ammunition, 430 grenades, 12 rocket-propelled grenade launchers, 12 DHSK machine guns, more than 50 SA-7 surface-to-air missiles, 2,000 electric detonators, 4,700 fuses, 106-mm cannons, anti-tank missiles, and 2 tons of Semtex.

To have access to plastic explosives is not enough, however, because effective use requires a certain amount of expertise. Would-be assassins failed in their attempt to kill France's President Charles de Gaulle in his car on September 8, 1961, because the 66 pounds of "plastic" used were not skillfully tamped into the ground beside the road. As a result, most of the charge did not explode. The Euskadi Ta Askatasuna (Basque Homeland and Freedom, or ETA) in Spain, however, was much more successful with its use of plastic explosives when a four-member cell assassinated Prime Minister Carrero Blanco in December 1973. Approximately 75–80 kilos of "Goma-2" were placed at the end of a tunnel along the horizontal part of the key and directly beneath the spot where Carrero Blanco's car would pass as he left church. To force his car over the correct spot and to cause his driver to reduce his speed, another car was double parked next to the location of the tunnel and explosives. . .on that fateful day at about 9:25 in the morning as Carrero Blanco's car passed over the explosives, the ETA commando, disguised as electricians working on cables along the street, detonated the charge. The force of the explosion threw the car over a five-story-high wall of the church and into the interior courtyard.

The training of saboteurs and explosives experts can therefore become an important function of the underground and often requires the support of external entities, such as the previously mentioned collaboration between the PIRA and the FARC. In other cases, an insurgent organization may gain knowledge via literature or online references. During World War II, the Soviet partisan network established a newspaper that instructed readers on (1) the placement of charges for optimum effect in sabotaging steel or wooden bridges; (2) the best points for sabotaging railway tracks, e.g., at bends or on high embankments; and (3) the alternative ways of sabotaging tracks, e.g., by using explosives or removing rivets from joints. Similar instructions were broadcast in a twice-daily, ten-minute program called "Course for Partisans," and the *Soviet Handbook for Partisans*, which contained detailed instructions, was distributed extensively.[5]

General Sabotage

An underground may conduct sabotage operations not only to hamper the enemy's war effort but also to encourage the populace to engage in general acts of destruction. Although the latter probably would not have much material effect, it could induce people to perform minor acts of sabotage and thereby link people more firmly to its cause.

In nonkinetic, general sabotage, the personal risk to the saboteur can be significantly reduced if the saboteur is able to remotely access the target area using information systems and/or cyber connections. The potential for this type of operation was demonstrated when, in 2008 and 2010, U.S. military computers were infected by corrupt thumb drives; the viruses impacted unclassified and classified networks as well as control systems at one base for Predator drones.[6,7] Because these viruses did not appear to be targeted at specific systems, it is assumed that their purpose was to cause a general disruption and widespread infection within the military networks.

Tactical Operations

To foster general sabotage, undergrounds often instruct the population in the use of certain sabotage techniques. These instructions are usually limited to the use of simple devices that do not require technical skills or elaborate equipment. For arson, the public may be instructed in the use of homemade incendiary grenades, or "Molotov cocktails." Fragmentary hand grenades can also be made with little trouble using an empty evaporated-milk can for the casing, a blasting cap, and metal pieces for filler and shrapnel.

There are many other easy-to-make devices for sabotage. Fires may be started by substituting incendiary solutions for nonvolatile fuels and by deliberately overloading machines. Mechanical interference may be produced by placing emery dust or sand in delicate bearings or by tossing pieces of scrap into moving mechanisms. Miscellaneous techniques for disrupting transportation include putting sugar into gasoline tanks, strewing nails on roads, blocking roads with stalled trucks and felled trees, and changing signs to misdirect traffic. In "passive" sabotage, the enemy is hampered when workers fail to lubricate machines, misplace spare parts, slow down production, practice absenteeism, etc.

Sabotage Intelligence

Effective sabotage missions are generally preceded by a tactical reconnaissance of the target and surrounding area. The specific information sought by this reconnaissance will vary with the choice of targets, but in general, it will include the exact location of the target and pertinent details such as structure, the number and positioning of guards (if any), the routine of the guards, the paths of access to the targets, the routes of escape, and areas for regrouping in case of dispersal. Before a sabotage mission proceeds to the target, the commander briefs the unit members on the plan of operations. Because of the decentralized nature of the underground organization, the underground does not usually jeopardize intelligence units by having them perform sabotage as well, for sabotage operations may draw attention to individuals and compromise their usefulness as intelligence agents.

ENDNOTES

[1] Joint Publication 3-24, *Counterinsurgency* (Washington, DC: Headquarters, Department of the Army, 2009), GL-6.

[2] Joint Publication 1-02, *Department of Defense Dictionary of Military and Associated Terms* (Washington, DC: Department of Defense, 2010 [as amended through January 15, 2012]), 318.

[3] Jerome Conley, "Revolutionary United Front (RUF), Sierra Leone" in *Casebook on Insurgency and Revolutionary Warfare, Volume II: 1962–2009*, ed. Chuck Crossett (Laurel, MD: The Johns Hopkins University Applied Physics Laboratory, 2009).

[4] JP 1-02, 291.

[5] ST 31-202, *The Underground* (Fort Bragg, NC: United States Army Institute for Military Assistance, 1978), 190.

[6] Andre Tartar, "Computer Virus Attacks U.S. Drones, But They Keep Droning On," *New York Magazine*, October 8, 2011, http://nymag.com/daily/intel/2011/10/computer_virus_attacks_us_dron.html.

[7] Willliam Matthews, "Pentagon to Allow Thumb Drives—Barely," *Marine Corps Times*, February 19, 2010, http://www.marinecorpstimes.com/news/2010/02/dn_021910_thumb2_w/.

GLOSSARY

Armed component: The visible element of a revolutionary movement organized to perform overt armed military and paramilitary operations using guerrilla, asymmetric, or conventional tactics.

Auxiliary: The support element of the irregular organization whose organization and operations are clandestine in nature and whose members do not openly indicate their sympathy or involvement with the irregular movement. Members of the auxiliary are more likely to be occasional participants of the insurgency with other full-time occupations.

Cell: The smallest organizational element of an underground formed around a specific process, capability, or activity. Cells are kept small for secrecy, and communication between cells is often limited to limit damage if any one cell is compromised.

Command and control: The exercise of authority and direction by a properly designated commander over assigned and attached forces in the accomplishment of the mission. Command and control functions are performed through an arrangement of personnel, equipment, communications, facilities, and procedures employed by a commander in planning, directing, coordinating, and controlling forces and operations in the accomplishment of the mission.

Compartmentalization: Establishment and management of an organization so that information about the personnel, internal organization, or activities of one component is made available to any other component only to the extent required for the performance of assigned duties.

Cyber attack: Computer-based attacks by state or non-state actors against public and private enterprises, including critical civilian infrastructure, that seek to exploit, deny, disrupt, or degrade networks, potentially resulting in physical damage and/or economic disruption. Attacks include theft or exploitation of data; disruption or denial of access or service that affects the availability of networks, information, or network-enabled resources; and destructive action including corruption, manipulation, or direct activity that threatens to destroy or degrade networks or connected systems.[a]

---

[a] Compiled from information included in *Department of Defense Strategy for Operating in Cyberspace* (July 2011): http://www.defense.gov/news/d20110714cyber.pdf.

Elite front organization: Organizational theory common to Communist doctrines that calls for establishing a "front" or "vanguard" to infiltrate existing liberation or independence movements and orchestrate the overthrow of the incumbent authority. The vanguard is theoretically the most ideologically advanced sector of society not prone to the "false consciousness" infecting mass society. The organizational theory was most clearly articulated by Lenin and adopted by the Bolshevik Party in Russia and has since been used by various theorists, including non-Communists, such as Sayyid Qutb in his influential book, *Milestones.*

Elite organization: Organizational theory that assumes a small, elite organization is capable of undermining the incumbent authority. Members are carefully targeted, vetted, and recruited.

Empirical sovereignty: Sovereignty of a political entity based on that entity's performance of governing functions such as the extension of force and the provision of social services. Political entities possessing empirical sovereignty are not necessarily officially considered states by the international community.

Extension of force: The ability of an actor, whether a state or a shadow government, to project force within its territory.

External shadow government: Government structure established by an underground outside the boundaries of the contested state.

General sabotage: General acts of destruction perpetrated by the populace and encouraged by the sabotage efforts of the underground. The material damage resulting from the acts is arguably negligible, but it does serve to link more people to the insurgents' cause.

Globalization: A concept used to describe the increasing interconnectedness of nations and peoples across the world, commonly understood to be fueled by the spread of neoliberal economic principles and technology.

Governance: Formal or informal methods of organizing political, social, and economic behavior. Governance is not synonymous with government—weak government institutions often lack the capacity to effectively direct and administer to the needs of citizens.

Hawala: An informal fund transfer system of moving money from one individual to another through the use of intermediaries and used especially by those in regions with limited financial infrastructure. The fees for the service are typically less than those found in the formal banking sector. Hawala has its origins in Islamic law.

Ideology: A set of beliefs that constitute one's goals, expectations and actions and form a comprehensive worldview. In insurgencies, a well-developed ideology serves the purpose of *unifying* disparate members of the movement, *organizing* actions around goals and shared values, and *justifying* actions that may include violence against countrymen.

Informal funds transfer (IFT)/alternative remittance systems: Networks of remitters that transfer money, often between countries, outside, or in the absence of, a formal banking system. Hawala is an example of an IFT system.

Insurgency: An organized movement aimed at the overthrow of a constituted government through use of subversion and armed conflict.

Internal shadow government: Government structure established by an underground within the boundaries of the contested state.

Juridical sovereignty: Sovereignty of a political entity predicated upon international recognition of statehood. Political entities possessing juridical sovereignty may be weak states that do not necessarily possess empirical sovereignty.

Legitimacy: The recognition by a population of a government's right and capacity to govern. Sources of legitimacy can be legal, hereditary, or religious or based on consent of the governed, endorsement by key influences, or other sources. Legitimacy tends to be undermined by poor governance, including unchecked corruption, crime, civil violence, lack of basic services, and economic deprivation.

Logistics: Activities associated with procuring, storing, and distributing supplies as well as maintenance, medical services, and transportation. Military supplies include food, water, general supplies, fuel and oil, building materials, ammunition, major end items like weapons and vehicles, medical supplies, and repair parts.

Mass organization: Organizational theory that assumes a large number of people are necessary to overcome the power of the incumbent authority and its instruments of force. Recruitment takes place on a region-wide scale.

Military intelligence: Intelligence providing valuable data about the enemy and the area of potential combat, such the number of enemy troops, their deployment, unit designations, the nature of their arms and equipment, the location of their supply depots, the placement of their minefields, and various other topographical features.

Parallel financial systems: Financial systems, especially those based on Islamic principles, operating in tandem with the primary Western system of institutions and regulations.

Political intelligence: Intelligence regarding pertinent information on regimes, regime activities, and regime supporters in critical countries.

Private military contractors: Private business entities that provide military and/or security services. Military and security services include armed guarding; protection of persons and objects; maintenance and operation of weapons systems; prisoner detention; and advice or training of local forces and security personnel.

Public component: The overt political component of an insurgent or revolutionary movement. Some insurgencies pursue both military and political strategies. At the termination of conflict, or occasionally during the conflict, the movement can transition to the sole legitimate government or form part of an existing government. Thus, the four spheres—armed component, underground, auxiliary, and public component—have a dynamic and evolving relationship changing in response to internal and external drivers. The public component's public position distinguishes it from the clandestine underground. However, it frequently overlaps with the underground in that the latter's functionality includes the management of propaganda and communications in general.

Radicalization: The process by which an individual, group, or mass of people undergo a transformation from participating in the political process via legal means to the use or support of violence for political purposes (radicalism).

Sabotage: An act or acts with intent to injure, interfere with, or obstruct the national defense of a country by willfully injuring, destroying or disrupting, or attempting to injure, destroy, or disrupt any national defense or war materiel, premises, or utilities, to include human, economic, cyber, and natural resources.

Sabotage intelligence: Intelligence providing information that aids in destructive attacks on critical and vulnerable infrastructure or personnel in an attempt to weaken government control and legitimacy.

Safe haven: Any space, whether physical, legal, financial, or cyber, that enables insurgent organizations to plan, organize, train, conduct operations, or rest with limited interference from enemy or counterinsurgent forces.

Safe house: An innocent-appearing house or premises established by an organization for the purpose of conducting clandestine or covert activity in relative security.

Selective sabotage: Kinetic or nonkinetic acts perpetrated by undergrounds in an attempt to incapacitate targets—whether related to transportation, information, the economy, or human resources, for example—critical to effective enemy military or government functioning. Tactical, rather than strategic, targets, such as bridges necessary for moving troops and supplies, are preferred because their destruction rapidly and decisively hampers enemy efforts.

Shadow government: A parallel governance structure established by an insurgent group that mimics the functions and attributes of the nation-state. Its functions include one or more of the following: extension of force, provision of social services, national identity and legitimacy, and revenue generation.

Social media: A set of Internet-based applications resulting from technological advances in communications technology that allows the creation and exchange of user-generated content. Social media supports informal, usually text-based communication in one-to-one, one-to-many, and many-to-many formats. Examples of three popular services that are considered social media are Facebook, Twitter, and YouTube.

Subversion: Actions designed to undermine the military, economic, psychological, or political strength or morale of a governing authority.

Underground: A clandestine organization established to operate in areas denied to the guerrilla forces or conduct operations not suitable for guerrilla forces.[1]

Ungoverned territory: Territory not effectively controlled by a state government.

Weak state: A state whose central government cannot effectively control or implement policies in the territory under its jurisdiction according to international borders and often suffers from a lack of legitimacy.

Zakat: Mandatory annual Islamic charity of approximately 2.5 percent of income from Muslim individuals, institutions, and companies. The charity is intended for the less fortunate but is also sometimes funneled to fund jihad.

ENDNOTE

[1] TC 18-01, *Special Forces Unconventional Warfare* (Washington, DC: Headquarters, Department of the Army, 2010).

www.ingramcontent.com/pod-product-compliance
Lightning Source LLC
Chambersburg PA
CBHW052111020426

42335CB00021B/2714